Also by George Gilder

Life After Television
Microcosm
The Spirit of Enterprise
Wealth and Poverty
Men & Marriage

SMTELECOSMTELECOSMTELECOSMTELECOSMTELECOSMTELECOSM**TELECOSM**TELECOSMTELECOSMTELECOSM

ELECOSMTELECOSMTELECOSMTELECOSMTELECOSMTELECOSMTELECOSMTELECOSMTELECOSMTELEC

MTELECOSMTELECOSMTELECOSMTELECOSMTELECOSMTELECOSMTELECOSMTELECOSM**TELECOSM**TEL

GEORGE GILDER

SMTELECOSMTELECOSMTELECOSMTELECOSMTELECOSMTELECOSM**TELECOSM**TELECOSMTELECOSMTELECOSM

ELECOSMTELECOSMTELECOSMTELECOSMTELECOSMTELECOSMTELECOSMTELECOSMTELECOSMTELEC

SMTELECOSMTELECOSMTELECOSMTELECOSMTELECOSMTELECOSMTELECOSMTELECOSM**TELECOSM**TEL

A TOUCHSTONE BOOK
PUBLISHED BY SIMON & SCHUSTER
New York London Toronto Sydney Singapore

ECOSMTELECOSMTELECOSMTELECOSMTELECOSMTELECOSMTELECOSMTELECOSMTELECOSMTELEC

ITELECOSMTELECOSMTELECOSMTELECOSMTELECOSMTELECOSMTELECOSMTELECOSMTELECOSMTELECOSMTEL

SMTELECOSMTELECOSMTELECOSMTELECOSMTELECOSMTELECOSMTELECOSMTELECOSMTELECOSMTE

TELECOSM

COSMTELECOSMTELECOSMTELECOSMTELECOSMTELECOSMTELECOSMTELECOSMTELECOSMTELECO

TELECOSMTELECOSMTELECOSMTELECOSMTELECOSMTELECOSMTELECOSMTELECOSMTELECOSMTELE

SMTELECOSMTELECOSMTELECOSMTELECOSMTELECOSMTELECOSMTELECOSMTELECOSMTELECOSMTEL

The World after Bandwidth Abundance

TOUCHSTONE
Rockefeller Center
1230 Avenue of the Americas
New York, NY 10020

First Touchstone Edition 2002
TOUCHSTONE and colophon are registered trademarks
of Simon & Schuster, Inc.
For information about special discounts for bulk purchases,
please contact Simon & Schuster Special Sales:
1-800-456-6798 or business@simonandschuster.com
Designed by Deirdre C. Amthor
Manufactured in the United States of America

10 9 8 7 6 5 4 3 2 1

The Library of Congress has cataloged
the Free Press edition as folows:

Gilder, George F.
Telecosm : how infinite bandwidth will revolutionize our world / George Gilder.
p. cm.
Includes index.
1. Telecommunication—Forecasting. I. Title.
TK5101 .G55 2000
384—dc21

00–055132

ISBN 0–684–80930–3
0–7432–0547–2 (Pbk)

For Josh, Anne Lee, and Max
with unchanging admiration and love

Contents

Prologue

Abundance and Scarcity

The computer age is over.

After a global run of thirty years, the PC revolution has stiffened into an establishment. So swiftly and subliminally did this silicon tide pass through the economy that like many experts you might have missed much of the motion until it stopped. Then you might have mistaken its dotage for dynamism. For it congealed so fast that its Mount Rushmore giants still walk and talk.

Contemplate the monumental frieze looming over the road ahead: Bill Gates of Microsoft, Steven Jobs of Apple, Gordon Moore and Andrew Grove of Intel. Still very much around, they beam from magazine covers, iconize companies, orate at Davos and Comdex, publish books (Grove even writes them). Jobs launches new products in "insanely great" new tints and hues. Their totems tower over their time: Windows 2000, the millennial operating system launched with 30 million lines of code and a record-setting two hundred thousand bugs; serried ranks of Pentium processors with scores of millions of transistors and insta-classical names, draining some 80 watts of power, enough to heat an igloo.

These Rushmore men, quick or dead, no longer shape the future. The trajectories of their companies are set in concrete source code, the

DNA of an epoch that is over. They are now retired to manage their portfolios, more and more sodden with deductible good deeds of planetary angst and posterity preening. The action is elsewhere. The action is in the telecosm.

Of course, the computer and the microchip remain enormously potent technologies. (So do the steel mill and the nuclear power plant.) Gordon Moore's law, which dictates a doubling of computer power, or a halving of its cost every eighteen months, is still in force. The process of ingraining intelligence into every aspect of our lives, mind into every machine, tool, or toy, continues at an accelerating pace. The displacement of matter by mind in the economy, already the most powerful economic event in recorded history—just now transpiring in economic data—has not yet even begun to end.

The computer era—the age of the microchip, which in a previous book I termed the *microcosm,*—is ending not because it has failed, or even because it has been fulfilled, but because the microcosm itself has given birth to a new era. It has enabled a new technology that is transforming culture, economics, and politics far more thoroughly than the computer age did.

The computer era is falling before the one technological force that could surpass in impact the computer's ability to process and create information. That is communication, which is more essential to our humanity than computing is. Communication is the way we weave together a personality, a family, a business, a nation, and a world. The *telecosm*—the world enabled and defined by new communications technology—will make human communication universal, instantaneous, unlimited in capacity, and at the margins free. In industry the word most commonly used for communications power is *bandwidth.* In the new economy, bandwidth replaces computer power as the driving force of technological advance. The telecosmic vision of nearly infinite wave-borne bandwidth does for communications what Moore's law did for computing: defines the direction of technological advance, the vectors of growth, the sweet spots for finance.

Like all the twentieth century's most crucial passages, this millennial transition has its roots in physics. While the industrial age emerged from a mastery of the masses and energies of Isaac Newton, the computer age sprang from a practical grasp of the particles and paradoxes of the quantum theory of Erwin Schroedinger, Werner Heisenberg, and

Albert Einstein. The heroes of the semiconductor industry took us step by step down quantum ladders toward the ultimate infinitesimal realms of atoms and ions. This path concludes in the quantum well of Heisenberg's uncertainty and reemerges in a coherent beam of waves.

In this path toward free computer cycles and bits, the transformative force was miniaturization—transistors multiplied first by the thousands, then by the millions, and tomorrow by the billions on a single sliver of silicon, enabling us to process, create, compute, and reconfigure information at a speed and scale previously unimaginable. Because the computer made the creation and manipulation of information the central activity of the economy, this era has also been known as the information age. We live in an information economy.

The great frustration of the computer era has been the difficulty of communicating the information that we are told has become our most precious resource. Information is power, but information that cannot be readily moved is gridlock on the World Wide Wait. Immobile information makes our businesses larger, more static and hierarchical than they need to be. It makes our economies less flexible, our jobs less fulfilling, our lives less luminous with opportunity.

The telecosm launches us beyond the fuzzy electrons and frozen pathways of the microcosm to a boundless realm of infinite undulations. Beyond the copper cages of existing communications, the telecosm dissolves the topography of old limits and brings technology into a boundless, elastic new universe, fashioned from incandescent oceans of bits on the electromagnetic spectrum.

At the heart of the telecosm are lasers pure enough to carve a sliver of light into thousands of usable frequencies, gossamer glass threads carrying millions of times more information even though they are thousands of times smaller than copper cables, fiber strands made of glass so pure that if it were a window you could see through seventy miles of it. The telecosm can even banish all the glass and unveil new cathedrals of light and air alone.

This is not futurism, for the science behind it is already history. The impact of the change, though, will exceed most of the dreams of technological futurists. Futurists falter because they belittle the power of religious paradigms, deeming them either too literal or too fantastic. Yet futures are apprehended only in the prophetic mode of the inspired historian.

The ability to communicate—readily, at great distances, in robes of light—is so crucial and coveted that in the Bible it is embodied only in angels. Distance is a fundamental premise of a material world. It fell not to the force of the telegraph, the telephone, the television, or the airplane. None of these achieve true action at a distance. Transmitting a few words, a few minutes of voice, even the few filmed spectacles that broadcasters deign to bounce around the globe, serves only to remind us how bound and gagged we are—how tied to the limits of time and space that angels traverse in an instant.

These gags and ties are now giving way. When anyone can transmit any amount of information, any picture, any experience, any opportunity to anyone or everyone, anywhere, at any time, instantaneously, without barriers of convenience or cost, the resulting transformation becomes a transfiguration. The powers it offers bring us back to the paradigms of paradise and its perils, prophets and their nemeses: infinite abundances and demonic scarcities.

The concept of infinitude challenges us all, even the mathematicians and technologists among us. But the central event in technology over the last decade is a growing awareness that the information bearing power of the electromagnetic spectrum—its bandwidth or range of frequencies and wavelengths available to carry signals—is not severely limited, as previously believed, but essentially infinite. From AM radio signals through microwaves to visible light—a band of frequencies millions of times the bandwidth of our 56 kilobit modems—the spectrum can carry usable signals. Nor is spectrum restricted to wireless bandwidth. Medium neutral, it can be distributed wherever there are coaxial cables, phone wires, or fiber-optic strands of glass.

An infinitude of potential bandwidth implies the endless multiplication of spectrum use and reuse. Cellular technologies have been created that allow the reuse of all available bandwidth in every cell, the sharing of cellular bandwidth among many users, and the proliferation of local cells through the deployment of more antennas.

The most powerful of all spectrum reuse technology is fiber optics. Every fiber-optic thread, the width of a human hair, can carry a thousand times more information on one path than all current wireless technologies put together. The basic measure of bandwidth is hertz or wave cycles per second. The bandwidth of currently used spectrum, running from AM radio to Direct Broadcast Satellite, comes to a total of some

25 billion hertz, or in scientific notation 25 times 10 to the 9th. The capacity of a fiber-optic cable is measured in petahertz—10 to the 15th waves per second.

These technologies of fiber optics, high-spectrum communication, and wireless networks, together comprise the telecosm. The telecosm makes bandwidth—information at enormous speed and almost infinite scale—the defining abundance of a new era, eclipsing even the still fantastic abundance of the computer age. It makes men into bandwidth angels.

Every new era is marked and measured by key abundances and scarcities. They shape the field of economics, the substance of business, the fabric of culture, the foundation of life. As Japanese futurist Taichi Sakaiya has written: "Survival dictates that human beings . . . develop an ethics and aesthetics that favor exploiting fully those resources that exist in abundance, and economizing on items that are in short supply." That is how we exist. We do not breathe xenon or eat platinum. Abundances and scarcities are not merely economic phenomena; they help weave the texture of life.

The defining abundances and scarcities do not transpire on ceremonial dates ushering one epoch in with trumpets and another out with violins. Eras overlap. The successes of one era spur and enable the successes of the next. The plenitude of the agricultural age loosed resources for the industrial revolution and the computer era. The mines, factories, nuclear plants, and oil fields of the industrial age are sustaining the metaphysical marvels of the age of information. The billion computers slated to throng the Internet over the next five years each will consume as much as a thousand kilowatt hours a year. Together with peripherals and hundreds of billions of embedded chips, Internet computing will use as much electricity as the entire U.S. economy does today, some three trillion kilowatt hours.

Fathoming a layered fabric of abundances and scarcities and finding those that mark a particular time is a challenge that has confounded many historians. Whether of food, fuel, land, minerals, or workers, scarcities are often salient; the politician's shoe pinches. Meanwhile the abundances which transcend or obviate the scarcities remain elusive or are as yet uninvented. Agrarian supremacists saw the industrial revolution as a source of oppression of workers in "dark satanic mills" and ignored the new abundances of housing, food, and clothing that lengthened lives by some fifteen years. Luddites and reactionaries can always claim that grain or land or heavy manufacturing are still the staff of life and source of real value, while the redeeming abundances of computer cycles or information are somehow spurious—not real output.

Reflecting the deceptive salience of scarcities, materialist superstitions often lead to what are called zero-sum assumptions of economic reality—a game in which a gain by one party necessitates a loss for another party. The gains and losses always add up to zero.

For most of human history, most people have believed that economics is essentially a zero-sum game—that scarcity will ultimately prevail over abundance. Whatever set of materials is central to production—whether land, food, oil, or the environment—will ultimately meet diminishing returns and exhaustion. Predicting a zero- or even negative-sum struggle for food, Pastor Malthus was the most famous exponent of the view that populations increase geometrically while agricultural output rises arithmetically. In the Malthusian view, food scarcity eventually chokes off growth. Karl Marx saw all economics ultimately reducing to a class struggle over ultimately scarce "means of production," with zero-sum assumptions and results. Throughout much of this century, dictators and warlords, from Hitler to Stalin, sought power and prosperity through a drive for *lebensraum* or the capture of neighboring territories, assumed to be the only source of new wealth in a world of scarcity.

As implied by the epigram, "Necessity is the mother of invention," scarcities, however, can yield abundances. Abundances and scarcities play out in a spiral of reciprocity, with each producing its opposite in the cycles of economic advance, but with abundance always finally prevailing in free economies.

The economists' focus on scarcity stems from the fact that shortages are measurable and end at zero. They constrain an economic model to produce a clearly calculable result, an identifiable choke point in the industrial circuitry. Abundances are incalculable and have no obvious cap. They tend to end in a near zero price and thus escape economics altogether. As the price declines and their role in the economy becomes more vast and vital, their role in economic analyses diminishes. When they are ubiquitous, like air and water, they are invisible—"externalities." Yet abundances are the driving force in all economic growth and change. In free economies, scarcities find their meaning chiefly in the abundances they engender and constrain.

Nations, companies, and individuals that exploit the "free" abundance (that is, the resource with the plummeting price) gain market share against all rivals. These groups come to define the very character of their age, whether an age of "steam" or "oil" or an age of "information." The political leaders who accommodate them become the prime movers in global affairs. Their countries pioneer and prosper.

In the agricultural age, land and manpower were tantamount to free. You wasted them to prevail in battle or commerce, to win wealth or political power. During the preindustrial era in America (from the founding through about 1825), the scarcity was horsepower and the abundance was land.

In the industrial age, physical force became as much as free. Fuel was a defining abundance. Horsepower—physical force, translated eventually into watts, or kilowatt-hours—abounded while land grew relatively scarce. We splurged on cheap horsepower—to clear farmland, to refine ores, to produce new farm equipment, to manufacture goods and capital gear, and to contrive armaments for capturing or defending land. And these activities defined the industrial paradigm.

Between 1660 and 1950, the cost of an effective kilowatt-hour dropped from thousands of dollars to some seven cents. It has since dropped to five and one half cents and continues to decline about 2 percent a year. In the industrial age, whether one were a tinker, a tailor, a naval captain, a candlestick maker, a general, or a farmer, one had to use steam power, coal power, oil power, and then electrical power in vast amounts in order to prevail economically. It was the defining abundance of the time and because of its centrality, it lasted a long time on top.

Over the last thirty years, we have been undergoing the computer revolution. In the era of the microcosm, transistors became asymptotically costless.

On a computer memory chip, the price of a transistor, with support circuits, dropped from some seven dollars to a few millionths of a cent. Declining at an annual rate of 68 percent, the price of a bit is now plunging close to a millionth of a cent as the billion transistor device—the gigachip—is introduced, on its way to an eventual price of under ten dollars. Dropping at an average of 48 percent per year, a MIPS (millions of instructions per second) of computer power that cost several million dollars in 1960 sells for less than a dollar today. Thirty-five years ago, a chip factory could produce a few score transistors a day. Today, a single production line in a microchip wafer fabrication facility ("fab") can produce some 1.6 trillion transistors in twenty-four hours. In 1999, wafer fabs produced some fifty thousand trillion transistors. We ended up wasting billions of them playing solitaire or controlling the environment in our cars or strumming guitars on music synthesizers.

In this microcosmic era, every nation, company, and individual—from the oldest professions to the newest, whether you are a policeman, a farmer, a general, an architect, a prostitute, a software engineer, a fitness coach, an oil geologist, or a pharmacist—you have to use transistors to succeed. Because transistors chiefly served to manipulate words and symbols in computers, the era has been renowned as the computer age. A radio station or a farm office, a police car or a plow, are all now congested with silicon. The transistors, each etched by the millions on chips the size of a thumbnail, are nearly as cheap as the beach sand of which they are made.

Since computers chiefly deal with information, many observers have defined the era as the age of information. But that has not exempted hard asset people from using this abundant technology. From Desert Storm to FedEx, from Wal-Mart to DreamWorks, from Ford to the NBA, from Century 21 to Merrill Lynch, the most successful ventures have exploited the abundance of cheap MIPS and bits. The harvest of the microcosm has come in the form of literally hundreds of billions of computers, each one more powerful than the "huge" mainframes of the 1960s, animating every appliance from a microwave oven to a permawave comb, from an airbag to a "hairball" (which is what Sun Microsystems' Scott McNealy ungenerously dubs Bill Gates's software).

Less noticed than the abundance of transistors but also important in shaping the technology of the era has been an implicit abundance of silicon area—the space available to accommodate the trillions of transistors. Filling up the multiplying expanses of computer backplanes and mother boards and daughter cards with increasing numbers of ever larger chips, silicon area has expanded dramatically. Every five years, the number of chips produced has approximately doubled, to some 285 billion units in 1999. With the average size of a chip rising some 50 percent during this span, total silicon area has been increasing 150 percent every half decade. Now at around 40 square kilometers of bare chips a year, the semiconductor industry could annually coat Silicon Valley itself, from San Jose to Palo Alto and across to Santa Cruz, with scores of layers of crystalline silicon meshed with microscopic aluminum and polysilicon wires and doped with enough exotic chemicals to be condemned as a toxic site by the Environmental Protection Agency.

In sum, the abundances of the microcosm have been transistors, power, and silicon area. An era's defining abundances find their meaning through its defining scarcities. In general, people use the abundances to relieve the scarcities. While in the macrocosm they substitute kilowatt-hours for horses and slaves, in the microcosm they use transistors to compensate for a shortage of human servants and broadband communications capacity. Think, for example, of the billions of chips devoted to making our telephone switches, modems, faxes, and "fast" Internet links function at all over narrowband telephone wires. It was in response to this basic problem that the transistor was invented in 1949 at AT&T's Bell Labs.

Just as scarcities create new abundances, abundances can create new scarcities. The availability of cheap food and housing, ushered in by the industrial era, caused a decline in the numbers of people willing to labor cheaply. The plethora of cheap fuel created a dearth of roads and a need for pollution controls. The more recent glut of transistors—and the colossal streams of bits they shaped and sent—has led to a shortage of communications capacity and technical manpower. While existing telephone bandwidth was ample for voice communications, it became suddenly scarce when faced with a global abundance of computers generating data at a rate of megabits per second. The canonical abundance of one era creates a canonical shortage for the next.

In the age of the telecosm, all the defining abundances of the computer era—ever cheaper power, transistors, and silicon area—are becoming relatively scarce. The most common digital devices will be cellphones and smart cards that cannot be simply plugged into the wall. These portable appliances must be powered by batteries or solar cells, technologies that double their efficiency over a span of decades rather than months. In cellphones and smart cards—increasingly performing an array of computer functions through speech recognition—power will be scarce.

Also scarce in these portable devices will be silicon area. Engineers must cram an ever increasing array of radio frequency, digital signal processor, and microprocessor functions within the constricted cavity and power budget of a handset. No longer can chips sprawl by the hundreds across PC backplanes and peripheral boards. The constraints of million-transistor designs dictate that cellphone computers will be based on single chip systems, sharply economizing on power and silicon, two defining abundances of the computer age.

Similarly, coupling lasers and other communications devices to a fiber optic thread with an 8-micron core, a tenth of the width of a human hair, imposes severe restriction of space and power. These constraints become more acute when the fiber is laid on the bottom of the ocean under a pressure of ten thousand pounds per square inch.

• • •

In the new millennium, all the defining abundances of the microcosm are becoming relatively scarce and expensive. This reversal is forcing a massive and drastic reorientation of the entire structure of the information economy. Every electronic system and infrastructure must be reformed to take advantage of the new canonical abundance. Measured by the expansion of Internet traffic, the price of bandwidth is decreasing and its availability is increasing by a factor dwarfing Moore's law. Bandwidth is demonstrably advancing at a doubling rate of at least four times the eighteen-month pace of the microcosm.

Visiting Lucent's giant fiber-optic manufacturing facilities in the Atlanta suburbs in late 1999, I saw evidence of a sharp further acceleration in the expansion of bandwidth. It is now practical to put a thousand wavelengths on a single fiber, ten billion bits of information per second on each wavelength, and as many as 864 fibers in each fiber cable. This adds up to a total of 8.6 petabits per second in a single fiber sheath.

Eight petabits per second is a thousand times the total average telecommunications traffic across the entire global infrastructure as recently as 1997. Adjusted for redundancies, eight petabits represented the total Internet traffic in 1999, per *month.*

Engulfed by this avalanche of bandwidth, we cannot readily measure it. But all the leaders in the world economy are changing course to ride the tides of light. A global economy designed to waste transistors, power, and silicon area—and conserve bandwidth above all—is breaking apart and reorganizing itself to waste bandwidth and conserve power, silicon area, and transistors.

Financiers are radically repositioning capital markets to take advantage of this transvaluation of values. With total Internet traffic and bandwidth doubling every three or four months, markets have to learn how to evaluate Internet companies now facing less than one tenth of one percent of the volume they can expect some five years hence.

Reduced to irrelevance are all the conceptual foundations of the computer age. A new economy is emerging, based on a new sphere of cornucopian radiance—reality unmassed and unmasked, leaving only the promethean light.

This book is my attempt to explain where it comes from (the science story), how it is taking over (the engineering story), who is fighting it, who will ride it to victory, and what it all means. I believe all of these components are necessary to understand the telecosmic revolution, though they are each rich and complex stories. Readers who do not have particular interest in the science may wish to skip Part I. Readers who are looking strictly for investment advice could make do with Parts III, IV, and the backmatter. General interest readers may enjoy the business stories of Parts III and IV, but will be most satisfied by Part V. But readers who go to the trouble of understanding such issues as why photonics beats electronics; why dumb networks beat smart ones (the stuff of Parts I and II), will be the most rewarded by the stories of the heroes and villains of our new age (Parts III and IV), and most able to ponder the questions of Part V.

TELECOSMTELECOSMTELECOSMTELECOSMTELECOSMTELECOSMTELECOSMTELECOSMT-

ELECOSMTELECOSMTELECOSMTELECOSMTELECOSMTELECOSMTELECOSMTELECOSMT-

ELECOSMTELECOSMTELECOSMTELECOSMTELECOSMTELECOSMTELECOSMTELECOSMTEL

ECOSMTELECOSMTELECOSMTELECOSMTELECOSMTELECOSMTELECOSMTELECOSMTELE-

COSMTELECOSMTELECOSMTELECOSMTELECOSMTELECOSMTELECOSMTELECOSMTELE-

COSMTELECOSMTELECOSMTELECOSMTELECOSMTELECOSMTELECOSMTELECOSMTELECO

PART ONE

SMTELECOSMTELECOSMTELECOSMTELECOSMTELECOSMTELECOSMTELECOSMTELE-

COSMTELECOSMTELECOSMTELECOSMTELECOSMTELECOSMTELECOSMTELECOSMTELE-

NEW LIGHT

COSMTELECOSMTELECOSMTELECOSMTELECOSMTELECOSMTELECOSMTELECOSMTELECO

SMTELECOSMTELECOSMTELECOSMTELECOSMTELECOSMTELECOSMTELECOSMTELE-

COSMTELECOSMTELECOSMTELECOSMTELECOSMTELECOSMTELECOSMTELECOSMTELE-

COSMTELECOSMTELECOSMTELECOSMTELECOSMTELECOSMTELECOSMTELECOSMTELECO

SMTELECOSMTELECOSMTELECOSMTELECOSMTELECOSMTELECOSMTELECOSMTELE-

COSMTELECOSMTELECOSMTELECOSMTELECOSMTELECOSMTELECOSMTELECOSMTELE-

COSMTELECOSMTELECOSMTELECOSMTELECOSMTELECOSMTELECOSMTELECOSMTELECO

SMTELECOSMTELECOSMTELECOSMTELECOSMTELECOSMTELECOSMTELECOSMTELE-

COSMTELECOSM TELECOSMTELECOSMTELECOSMTELECOSMTELECOSMTELECOSMTELE-

COSMTELECOSMTELECOSMTELECOSMTELECOSMTELECOSMTELECOSMTELECOSMTELECO

SMTELECOSMTELECOSMTELECOSMTELECOSMTELECOSMTELECOSMTELECOSMTELE-

COSMTELECOSMTELECOSMTELECOSMTELECOSMTELECOSMTELECOSMTELECOSMTELE-

COSMTELECOSMTELECOSMTELECOSMTELECOSMTELECOSMTELECOSMTELECOSMTELECO

SMTELECOSMTELECOSMTELECOSMTELECOSMTELECOSMTELECOSMTELECOSMTELE-

Chapter 1

Maxwell's Rainbow

"Nothing is too wonderful to be true."
—James Clerk Maxwell, discoverer of electromagnetism

"Too much of a good thing can be wonderful."
—Mae West

The supreme abundance of the telecosm is the electromagnetic spectrum, embracing all the universe of vibrating electrical and magnetic fields, from power line pulses through light beams to cosmic rays. The scarcity that unlocks this abundance is the supreme scarcity in physical science: the absolute minimum time it takes to form an electromagnetic wave of a particular length. Set by the permeability of free space, this minimal span determines the speed of light.

The discovery of electromagnetism, and its taming in a mathematical system, was the paramount achievement of the nineteenth century and the first step into the telecosm. The man who did it was the great Scottish physicist James Clerk Maxwell. In his honor, we will call the spectrum Maxwell's rainbow. Today most of world business in one way or another is pursuing the pot of gold at the end of it.

Arriving at the profound and surprising insight that all physical phenomena, from images and energies to chemical and solid bodies, are built on oscillation, Maxwell embarked on a science of shaking. For roughly a hundred and fifty years, this improbable topic has animated all physics. Another word for oscillation is temperature. Without the oscillations, the mostly empty matter of the universe would collapse in on itself. In theory, you can make the shaking stop, but only by making things cold indeed—273 degrees below zero Celsius, or zero Kelvin. So far unreachable even in laboratories, it is the temperature of the universe's heat death.

When things oscillate, they make waves, and in that magic moment the possibility of the telecosm is born.

Maxwell's genius was to realize that all waves are mathematically identical, and can be arrayed along a continuum known as the spectrum. The unity of the spectrum makes possible the ubiquity and interoperability of communications systems and thus enables the unification of the world economy in the new era.

The light your eyes can see is only a tiny slice of the range of "colors" that actually exist or can be created. They run from the background rumble of the universe at the low, or "dark" end, to shrieking gamma rays that can penetrate a planet at the high "bright" end. Each wavelength has its own distinct characteristics—some are better at transmitting raw power, others for traveling long distances, others for carrying digital bits.

Slices of Maxwell's rainbow form the core of virtually every significant modern technology: 60-hertz household power cords and three kilohertz (thousand-cycle) telephones; 700 megahertz (*mega* is million) Pentium PCs; two gigahertz (billion) cellular phones and 200 terahertz (trillion) fiber-optic cables. The neurons in your brain, for their part, hum along at barely a kilohertz; thank the Lord for parallel processing. Dental X rays, at the other extreme, top a petahertz—a thousand trillion cycles per second. The potential number of frequencies is literally infinite, limited only by how finely your technology can parse the rainbow.

Maxwell's theory informed his several immense tomes on electromagnetism. The fruit of a promethean life ended by cancer at age forty-eight, his work empowered titans such as Erwin Schroedinger, Hendrik Lorentz, Albert Einstein, and Richard Feynman to create the edifice of twentieth-century quantum and post-quantum physics.

As much as pure scientists hate the idea, however, it is engineers and entrepreneurs who finally ratify their work. Until theory is embodied in a device, it is really not physics but metaphysics. Newton's ideas burst forth as the industrial revolution. Quantum theory triumphed unimpeachably in the atomic bomb and the microchip. In contrast to the intriguing perplexities of particle physics—Einstein's relativity, Murray Gell-Mann's quarks, Richard Feynman's quantum electrodynamics, Stephen Weinberg's grand unification, Schwartz's karass of superstrings—Maxwell's rainbow may seem child's play. But as we approach the twenty-first century, the spectrum's infinite spread of capabilities is history's driving force.

Maxwell had transformed the mindscape of metaphor and analogy by which human beings grasp reality. For Newton's medley of massy and impenetrable materials, he substituted a noosphere of undulatory energies. And woven uniquely into the warp of nature was the resonating speed of light. As Maxwell and others discovered, the speed of light is a basic constant in our universe—no matter the speed of the observor or the medium. Frequencies and wavelengths may change, but light speed delay—the time it takes to propagate an electromagnetic wave—never changes.

As we will see, light speed is both the crucial enabler and limit of the telecosm. Without it, radiation would be chaotic and uncommunicative. It would be noise that could not bear a signal. Yet communication can never exceed this speed, a fact that will keep us forever distant from other planets and even from ourselves.

There are no practical limits to the spectrum's range of possible wavelengths and frequencies. Nor is the spectrum expressed only by the physics of electromagnetic waves. Spectral frequencies translate into temperatures, into atomic signatures, and into photon energies.

Let the action begin by beating on a drum at a rate of once each second: one hertz. Translating these drumming "phonons" into electromagnetic form, a one-hertz frequency would command a theoretical wavelength of three hundred million meters. Applied to a single photon, its energy in electron volts would be Planck's quantum constant—6.63 times 10 to the minus 34th power, close to "Johnson noise," the background chill of the cosmos. Slowly accelerate the drumming to the fast be-bop rattle of a Max Roach or Buddy Rich, perhaps 16 beats per second. That is 16 hertz, around one fourth of the rate of an electrical power station. Suppose that your drumming skills are superhuman, moving at 3,000 beats per second; you are transferring the same number of oscillations that can be carried by a telephone wire. At some 30,000 hertz you have broken the sound barrier because you are sending out wave crests faster than they can be heard.

Nonetheless, you remain near the very bottom of the electromagnetic spectrum. At the other extreme are gamma rays, creatures of cosmic explosions and giant particle accelerators, a frequency of 10 to the 24th hertz. Their wavelength, 10 to the minus 22 meters, is small enough to get lost in an atom. Between Johnson noise and gamma rays is the telecosm, the gigantic span that Maxwell bridged with his mind, most of it now open to human use.

Above 14 gigahertz—at wavelengths running from the millimeters of microwaves down to the nanometers of visible light—is the new frontier of the millenium, empires of air and fiber that command some *fifty thousand times* more communications potential than all the lower frequencies we now use put together. A purely human invention, they provide the key arena of economic activity for the new century.

To put this huge span of frequencies in perspective, a factor of some 10 to the 25th stands between the lengths of the longest and shortest known forms of electromagnetic waves. As molecular biologist Michael Denton has observed: "A pile of ten to the twenty-fifth playing cards would make a stack stretching halfway across the observable universe." Seventy percent of the sun's light and heat occupies the band between near-ultraviolet and near-infrared—the width of the edge of just one playing card in Denton's cosmic stack. This little sliver of the spectrum providentially sustains life. Maxwell opened the rest of it up for human use: the telecosm.

Chapter 2

The Imperial Science

"The whole domain of Optics is now annexed to electricity, which has thus become an imperial science."

—Oliver Lodge, 1889

Maxwell gave us his rainbow, an infinite span of regular radiance that can reach around the globe and into space. *Infinite,* it could enable a global network rather than a spectronic Babel, a worldwide web of bandwidth abundance rather than a labyrinth of local loops. *Regular,* it could bear a measurable modulation, a deliberate distortion detectable at a distance, a stream of signals rather than a rush of noise. *Radiant,* it consisted of patterns of energy that could be enhanced and harbored, amplified and resonated through the media of the telecosm.

In the wake of gravity and thermodynamics, Maxwell's electromagnetic theories seemed the consummation of physics. The very sun itself—at once gravitational, thermodynamic, and electromagnetic—never set on the empire of classical theory. On the imposing pile of monumental nineteenth-century legacies, science, it seemed, at last could sleep. But how ever elegant and coherent, this science could not sustain a unified technology of human communications.

Underneath the towering matrices of the math lay infinitesimal and altogether impalpable to ordinary humans, a pea. The pea was the packeted energy allotment commonly known as the quantum. It was the smallest pea ever to disrupt the slumber of a physicist—6.65 times 10 to the minus 34th joule seconds—but to a few preternaturally sensitive thinkers it was big enough. In time it would give birth to the photon, the smallest glimmer of light, measured by multiplying that little pea times the frequency of the light. Granulated in photons, Maxwell's rainbow became a manageable radiance that could endow a global communications network.

Multiplied by frequencies of millions to trillions in telecom devices, these quanta could become streams of photons. Each with an indelible frequency signature, the photons could add up to bits that could be sorted and sent in accordance with their wavelength. Governed by the laws of interference, they could be enhanced by aligning their wave crests in phase, to make a laser, or deleted when the crests were in an opposite phase. Thus by manipulating the paths of the photons, engineers could combine them for transmission, or separate them by dividing the light like a prism into its constituent colors or wavelengths. They could, in short, manipulate Maxwell's rainbow in order to convey information.

It was Albert Einstein who, in 1905, first felt the pea beneath his mattresses. In a famous paper, the twenty-six-year-old patent clerk boldly walked the "planck" of the quantum where Max Planck himself had feared to tread. To a reluctant world of wave theorists, Einstein introduced the photon.

He insisted that quanta were everywhere, even in radiation through space. Rather than a continuous spread of energies, as Maxwell had hypothesized, or even a manifestation of statistical probabilities, as gas theorist Ludwig Boltzman had proposed, Einstein asserted a radically new model for electromagnetic radiation. Light consisted of discrete photons. Far from being in any sense continuous, radiation was emitted at high temperature only in small bands separated by enormous gaps.

Four years later, in a speech published in *Physikalische Zeitschrift* in 1909, he summed up his argument before an audience that included Planck himself. Making a further point crucial to the emergence of telecosmic technology, Einstein used quantum theory to explain that light propagation could be reversed. Maxwell's theory had implied that waves propagated irreversibly outward toward infinity and thus could not be captured by a photoreceptor or atom at a long distance.

Einstein cited experiments with cathode rays and X rays that demonstrated such "essentially inverse processes" as he put it. These inverse processes are now crucial to lasers, photodetectors, light-emitting diodes, and other devices that make fiber optics and photonics technology possible.

Sitting in Einstein's audience, Planck rose up to object. "I am almost astonished," he said, "that there did not arise more opposition to this." Planck pointed out that for quanta to show the demonstrable interference effects that constitute the very definition of waves in space, they "would have to have a spacial extension of hundreds of thousands of wavelengths." Planck was understandably rejecting the now familiar quantum paradox of particles manifesting wavelike interference patterns or waves emerging in particulate concentrations.

Einstein, though, was some twenty years ahead of his audience. Rather than discarding Maxwell's equations, he proposed to integrate classical wave theory with quantum theory much as scientists do today. He noted that a photon behaved in some ways like an electron, in some ways not. "The field as produced by atomistic electric particles is not very essentially distinguished" from the field associated with a "particle" of light.

Einstein was essentially predicting the next twenty years of quantum electrodynamics—the work of Louis de Broglie, Paul Dirac, Werner Heisenberg, and Erwin Schroedinger—which would describe all quantum entities in terms of a wave and particle duality. By that time, Einstein himself would be refining his theories of relativity, which in their assumptions of continuity were ironically incompatible with his quantum theory.

For the purposes of the creation of the specific devices of the telecosm, however, Einstein's pinnacle came in 1917. In calculations of that year, Einstein offered what was perhaps his most fertile telecosmic idea: the stimulated emission of radiation. Now embodied in the *laser* (an acronym for light amplification through the stimulated emission of radiation), Einstein's concept relied on Niels Bohr's atomic model. This model hypothesizes that the electrons in an atom occupy a limited set of discrete energy levels or orbits. When an electron moves up or down, from one energy level to another, it either emits or absorbs a photon of a definite frequency.

These reciprocal processes manifest themselves in a fluorescent light, which is a glass tube filled with neon, mercury, or sodium gas which lights up when suffused by an electrical current (that is a stream of electrons). They excite the atoms in the gas to a higher energy state. As the gas atoms tumble back down to a lower energy level, they emit ultraviolet photons that excite the phosphorescent chemical coating of the lamp.

Einstein's vision of stimulated light differentiates it clearly from all other light we see. All the light in nature—from the sun, from stars, from electric lightbulbs—is *spontaneous* light. Atoms and molecules excited to a higher energy level by heat or some other electromagnetic source spontaneously release their excess energy by emitting photons—discrete packets of energy—as they drop down to lower energy levels. Einstein's enhanced emissions would have the property of *coherence:* its waves all the same length, lined up in the same phases of crests and troughs. The light would constitute a beam of powerful intensity because it would contain the amplitude of the original photons multiplied exponentially.

He realized that on many occasions a photon would strike an electron already in an excited state produced by a photon. In this situation, the electron would be "full"; it could not absorb the new photon. Therefore, Einstein proposed, the electron will emit a separate but identical photon moving in exactly the same direction as the photon that struck it.

Einstein stopped short of inventing the laser. However, multiplying this Einstein effect many times produces what is termed a *population inversion,* with millions of electrons pushed at once to a higher state. Bombarded by photons from other electrons similarly excited in a chain reaction, they end up as a polarized coherent stream of light. This laser action is a key foundation of fiber optics and photonics.

That was stimulated emission as Einstein conceived it. And also stimulated emission as it works in nature in the phenomenon known as the aurora borealis or Northern Lights, which in effect are a vast oxygen laser, excited by unusually intense discharges from the sun. Stimulated, coherent light defies the forces of entropy and equilibrium that tend to pull atoms to their lowest available energy state. The opposite of lasing is laziness—the principle of least action—a fundamental law of nature discovered by the nineteenth-century Irish physicist William Hamilton. As Hamilton noticed, most particles, most of the time, are inert. To stimulate light, by contrast, you must pump up the crowds—pump electrons into an excited state faster than they tumble to ground, and then watch the transcendent light pour forth into the telecosm.

Chapter 3

Enter the Laser

It would be a gas. An excited gas. Perhaps a "noble" gas. You could create a new device with many practical uses and at the same time fulfill Einstein's vision of stimulated light. They have awarded Nobel prizes for less.

At Bell Laboratories in the middle 1950s, it was Charles H. Townes's luminous dream. Look. . . . nature does it too, more or less. In Nobel's Stockholm, for example, such glowing vapors are common. Take an aurora borealis. It dawns like a laser of oxygen gas excited by charged particles from storms in the sun, pulled to the poles of the earth's magnetic field and briefly held in luminous suspense. Northern lights. As the electrons tumble down the steps of the quantum, and the king enters with the glittering medal in the mirrored room, the oxygen emits streaks of greenish and reddish photons on the horizon in the night. For a moment, the atoms in the sky have assumed a metastable state, a population inversion, where the majority are lifted to higher quantum energies and a chain reaction streams forth. In this condition, when the Nobel crowd rises to its feet, inverted as one, and you think the applause will never end, trapped and echoing across the mirrored chamber, the aurora lasts about a second, long enough for the world to see the coherent glow around your balding pate, a crown of northern light more enduring than the king's.

• • •

For more than three decades, Einstein's idea, voiced in 1917, stood out as a challenge to technologists. One possible route to making it work lay with gases, a controlled version of the aurora. With artful engineering, an enclosed cavity could theoretically produce the necessary population inversion, with excited electrons stimulated to create intensifying streams of photons. Bounce the resulting waves back and forth between mirrors, then open a small hole, and—like a light-filled flute—a magically pure tone streams forth.

For another metaphor, take a set of waves on the surface of a pond, spreading from a fallen stone. Their power diminishes according to the inverse square of their radii, just as ordinary light waves do from a bulb. But confine them to a bathtub and they will rock back and forth, piling up until they overflow the sides. As they rock they will impart rhythmic power to other waves, stimulating even more overflow. They will resonate and then propagate exponentially.

Combine those ideas and you have a laser. Bounce light waves between those mirrors and only synchronized frequencies will emerge, a beam of *collimated*—perfectly parallelized—exquisitely *monochromatic* light. Make one of the mirrors partly transparent and just the right waves will flow through in a coherent stream.

The idea of stimulated light makes sane men dream wild dreams—of laser-based fusion energy, simulating the fires of the sun; of light punching holes in metal, cutting cloth, and gauging the distance to the moon down to the millimeter. Stimulated light can work its wonders at infinitesimal powers inside microchips, or melt a mote off a retina. Bathed in the 77-degree Kelvin chill of liquid helium, a low-frequency laser device can—and did—detect the background radiation from the big bang at the beginning of time, winning the Nobel prize for Arno Penzias and Robert Wilson. Lasers can perform miracles. But to understand how all this can happen, you have to work with the quantum paradox—a feat that long defied some of the world's greatest minds.

The first scientist to harness stimulated energy into a working device was Charles H. Townes of Columbia University. A learned physicist who also worked for Bell Labs in Murray Hill, New Jersey, Townes had coauthored a definitive tome on microwave spectroscopy with his Bell protègè and brother-in-law, Arthur Schlawlow. Townes, no tinkerer, prefered to theoretically model the full behavior of a system before trying to build a working prototype. By 1955, he was able to do that for a gas-based version of what would later be known as a laser. A decade later, it would help win him a Nobel prize.

It was not just because he had written the book on them that Townes began with microwaves. A maser, as he called it—for microwave amplification through the stimulated emission of radiation—would also be less problematic to build than a device using visible light. As Maxwell himself could have predicted, higher frequencies require both greater stimulating power and a smaller generating chamber. One obvious answer to that would lie in the microcosm. But this was the 1950s and most scientists—Townes included—preferred elegant tubes and gases to their day's still crude semiconductor devices.

Even after Townes had a working device, using microwaves, coherent light seemed too miraculous for much of the physics establishment to swallow. In 1956, two midcentury lords of science, Niels Bohr and John Von Neumann, visited Townes's lab at Columbia. Bohr was the reigning dean of quantum theory. Von Neumann, a polymathic genius, had just laid out the architecture of the modern digital computer. A pure beam of perfectly aligned photons, the two giants told Townes, was quite impossible.

Coherence, Bohr and Von Neumann argued, implied regulating lightwaves with a precision that Heisenberg's famed uncertainty principle did not allow—if the location of particles could not be predicted, they could hardly be marshalled into orderly rows, marching lockstep through space. What the two great men missed was the power of resonance, and the perfection of energy waves. Just as resonance perfectly lines up the quantum levels of atoms, resonance can line up photonic waves, and make them rock spontaneously to a rising rhythm. As Townes wrote after the visit from on high, "To build a laser, you have to be able to grasp both horns of the quantum dilemma at once. Emphasis on the photon aspect of light deflected some physicists from coherent amplification."

Plenty of others needed no convincing. Around the time that Townes and Schlawlow were publishing their results and the implications, a perennial graduate student at Columbia, Gordon Gould, was filling notebooks with his own ideas about light stimulation. Gould, a former student of Townes, had been invited to his laboratory in late October 1957, to discuss an idea Townes had for using thallium lamps—then still under top-secret official wraps—for pumping stimulated light. An avowed Marxist at the height of the cold war, Gould was barred from independent access to such militarily sensitive information. But two weeks later his notebooks were full of ideas—and notarized by his Uncle Jack in the Bronx. They read in places like a *Popular Mechanics* article, a stream of consciousness catalog of stimulated gases and hypothetical devices, including everything from metal drilling and welding to heavy-water nuclear fusion.

If the Nobel was Townes's dream, Gould's notebooks were Townes's nightmare. Their owner had yet even to finish his thesis, let alone build anything that could singe a Kleenex. But while Townes and Schawlow were publishing papers on what they innocently called "optical masers," Gould was busily applying for a patent, using his notebooks for proof and calling the device a "laser." A nightmare indeed. Couldn't he get it through his head that a maser was a laser? The laser had already been invented!

Although Gould and his team did not build a working device for years, his notebooks and talent at coining names made him the financial winner among the early laser pioneers. In later decades, just as the laser industry exploded into a $600 million bonanza, he won court decisions granting him a stream of royalties for optically pumped lasers, gas lasers, and drilling, cutting, and welding with lasers—a fortune that ultimately approached $50 million. Not bad for a former Marxist perennial grad student. But Gould's vision was wrong in terms of the telecosm, where microwaves, radio waves, X rays, gamma rays, and visible light were concerned—all are, as Townes insisted, indivisibly part of Maxwell's rainbow. At least Townes got his Nobel prize.

Ironically, neither Townes nor Gould built what everyone agrees is the first working device to stimulate visible light. Both Townes's team at Bell and Gould's new Long Island based company, TGR, had thrown themselves down the largely blind alley of gas lasers, fixing on gasified potassium as a medium. This gas had suitable spectronic characteristics, but also an unfortunate tendency to blow up during the process and blacken the lasing chamber. Only in 1962, when he turned to a mixture of neon and helium, did Townes eventually produce a working laser.

That was two years too late. In May 1960, only four years after

Bohr and Von Neumann assured Townes of its impossibility, the first laser was built, and it was solid-state—no messy gas-filled tube was needed to generate its bright red burst of photons. At its heart it had a ruby, which "everyone," including Townes and Schlawlow, "knew" wouldn't work. Like any complex crystal, rubies were harder to model than a gas, made up of free-floating atoms. And that same complexity—coupled with the tiny size of silicon chips—was deemed an overwhelming bar to building miniaturized semiconductor lasers.

Far away at Hughes Research Labs in Culver City, California, it was Theodore Maiman, an ex-Stanford engineer, who built that first device. Maiman had taken another tack: he had reexamined the evidence on ruby photo-efficiency and discovered that the prevailing calculations were wildly wrong. Instead of one-percent efficiency at turning energy into light, the real figure turned out to be closer to a stunning 70 percent—a number he eventually was able to push to near one hundred.

Using a pink ruby cylinder recycled from one of his earlier masers, Maiman contrived a lasing chamber less than an inch long and half as wide, with both ends silvered to act as mirrors. At one end was a tiny hole to emit the coherent light. Spiraling around it all was a helical photographic flash lamp, to impart the energy for stimulated emission. The silvered mirrors were not perfect, but the simple little device did what it was supposed to: emit a brief flash of coherent light.

Within a month, Maiman submitted his amazing findings to the distinguished journal *Physical Review Letters,* under the title "Optical Maser Action in Ruby." But instead of applause and consternation from astounded friends and rivals, there was silence. Maiman had not used Gould's controversial new name, laser, for his device; and the journal's editors, thinking they were getting yet another maser article, turned it down. Only after Schlawlow and his team at Bell Labs verified his work was Maiman assured his historic pinnacle as creator of the first working laser, or, in deference to Townes, "optical maser."

It was, of course, Gould's more striking name that stuck. And in time, Maiman's little device grew into a significant business for Hughes. But in telecosmic terms, a more crucial breakthrough came two years later: lasers compressed inside a microchip. The tiny expanse of a semiconductor p-n junction—the interface between positively and negatively charged regions on a chip—ultimately proved to be the most efficient and useful of all lasers. Moving from large gaseous tubes to modest ruby crystals and finally to infinitesimal semiconductors, the smaller the laser, the more effective and powerful it would become. *The scarcer the room for the device, the more abundant its output.*

Physics simply could not grasp the complexities of solid crystals. Just as conventional scientific wisdom had deemed Maiman's rubies too inefficient to lase, physicists initially estimated the photo efficiency of semiconductors at a vanishingly small 10 to the minus 40—point 42 zeros—percent. But experiments soon started pushing the figure upward. As it transpired, semiconductor p-n junctions convert electrical currents into photons at an efficiency close to 100 percent. Nothing turned out to be so effective in creating a flood of photonic energy as a mass of negatively charged "male" electrons falling into an array of positive "female" holes. It was easier to create a population inversion by using semiconductor bandgap engineering than by manipulating gasses, inch-long crystals, and vacuum tubes in the macrocosm.

The essence of laser science quickly become molecular manipulation, the central art of the microcosm. Both in Leningrad and at Bell Labs in the late 1960s, engineers contrived heterojunctions, combining multiple elements to confine the reactions in a lasing device. Zhores Alferov and his team developed the first room-temperature laser, but Soviet politics stifled further development. Mort Panish and Izuo Hayashi at Bell Labs also made dramatic progress, but the lasers lasted mere minutes rather than the millions of hours needed for telecom devices. Barney de Loach of Bell finally succeeded in contriving a room-temperature laser that lasted. It was a heterojunction made of gallium arsenide, aluminum, indium phosphide, and other materials which combine to form energy steps down which cascading photons resonantly tumble. Quantum well devices, the smallest and most efficient lasers in widespread use today, reduce this process to the size of a single electron wavelength. Costing barely three dollars a piece in 1999, they are manufactured by the million for use in CD players.

In a climax to the laser's microcosmic evolution, Jerome Faist and Federico Capasso at Lucent's Bell Labs in the late 1990s moved the lasing area down the level of individual quanta, generating photons from an array of twenty-five separate quantum wells. Each successive stage is set at a lower energy level than the one before, like a quantum staircase. Carried from one well to the next by transmission channels, a passing electron emits not one photon, but *twenty-five.* Unlike most lasers, this still-experimental device can propagate coherent light through all the visible and infrared bands of Maxwell's rainbow.

As Gordon Gould's fevered notebooks predicted, lasers have emerged as weapons, as antiweapon missile shields, as cancer treatments, as spectroscopic tools, theatrical props, and fusion energy sources. But their most important use was scarcely mentioned in the speculations of most laser pioneers: fiber-optic communications.

Even well into the 1960s, most scientists who thought at all about laser-borne information streams foresaw short-range links through the air or down waveguides—hollow tubes. Pondering the problem of atmospheric distortion, a few envisaged connections between satellites in orbit. But glass remained the mostly obscure province of CIA spooks and people like Will Hicks, who was coinventor of a basic device of fiber optics and my initial guide to the telecosm. Just as the discovery of 100-percent photonic efficiency in p-n junctions, fiber optics for telecommunications would burst on the scene as an unexpected miracle.

Chapter 4

The Light-Speed Limit

Let there be light, says the Bible.

All the firmaments of technology, all our computers and networks, are built with light, of light, and for light, to hasten its spread around the world. Light glows on the telecosm's periphery; it shines at its core; it illumines its webs and its links. From Newton, Maxwell, and Einstein to Richard Feynman and Charles Townes, the more men have gazed at light, the more it turns out to be a phenomenon utterly different from anything else. And yet everything else—every atom and every molecule—is fraught with its oscillating intensity. In a very real sense, the world *is* light.

In all its forms and frequencies, light sets boundaries, just as much in the tiny expanse of the microcosm as in the telecosm's infinite domains. Its paradoxes dictate topographies, network architectures, the shape and size of machines. The further we plunge ahead, the more speed of light becomes the defining scarcity, the key constraint. We face once again the great predicament of the outset of the twentieth century, when physicists first faced the full meaning of an absolute speed of light.

By many measures, Isaac Newton was the greatest scientist of all time—the man who most radically, accurately, and consequentially reshaped the conceptions of reality held by human minds. But he also made two huge mistakes, or oversights. In part because a seventeenth-century scientist had little more to work with than what his own eyes could see, he assumed that matter was solid, made of hard, unbreakable atoms. And for the purposes of his *Principia,* he assumed that light's speed was infinite—that it moved instantaneously through space.

Quantum theory took care of the first error: matter is not solid, as Newton assumed, but consists of atoms as empty in proportion to the size of their nuclei as the solar system is empty in proportion to the sun. Within these atoms, at the heart of everything humans see and experience, Newton's laws turned out to be irrelevant and wrong.

Maxwell capsized the other Newtonian misconception, instantaneous light, by coming up with equations that yielded an indisputably finite velocity. But at the time, the late nineteenth century, no one could fully gauge the implications.

The problem was partly Maxwell's own. His dogged belief in luminiferous ether—a mysterious, light-bearing medium that filled the universe and explained wave theory—suggested that light would move faster from a source moving through the ether than from a source at rest. Common sense concurred: Light from your headlights *should* move faster, or so it would seem, if you are speeding toward a photodetector than if you are backing away. But rest assured, it doesn't.

It would be nearly a century before scientific instruments improved sufficiently to measure light's speed directly in the lab. But comparing its speed from two different sources, one in motion, the other not, was feasible—provided you could find an object moving fast enough to impart a measurable difference in light's vast velocity. That high-speed platform turned out to be literally under the experimenter's feet: Earth itself, orbiting around the Sun at eighteen miles a second, a full ten thousandth of light's speed.

In 1887, two American experimenters, Albert Michelson and Edward Morley, tested the idea that the earth's orbital velocity affected the speed of light. In an ingenious experiment, they split a beam of light and sent it in perpendicular directions to two identical mirrors. When the beams bounced back, they converged in a device—called an interferometer—invented by Michelson for the purpose. If the speed of light was affected by the earth's movement through the ether—as most scientists of the day thought it would be—then the beam moving parallel to the earth's orbit would move faster than the one traveling perpendicular. The expected result—a shift of some 40 percent of the chosen light's wavelength—would produce an identifiable interference pattern at the spot where the two converged.

Meticulous scientists, Michelson and Morley performed the experiment twice, rotating the whole apparatus 90 degrees after the first attempt, to compensate for possible discrepancies between the two light paths. But the results showed absolutely no impact from the earth's orbital velocity. In 1958, laser inventor Charles H. Townes confirmed the Michelson–Morley results at Bell Labs using high-frequency radio waves, another form of light.

By defining all electromagnetic radiation by a ratio that involved this velocity, c, Maxwell had made the light-speed limit inherent in the phenomenon itself. Nuclear experiments in 1937, using Einstein's canonical 1905 equation, $e=mc^2$, further established the total autonomy of light. The experiments compared the masses (m) before and after a nuclear reaction and measured the amount of energy (e) released in the reactor. C^2 remained and it equaled the immense ratio of the energy over the mass. Since both the masses and the energies were measurable or known quantities, the equation could be solved for c, assumed to be unknown. That calculation, involving no conventional speed measurements at all, yielded the velocity of light to an accuracy of half a percent. Light speed is an absolute. It is intrinsic to the resonant fabric of the universe.

In the microcosm, light provides crucial stability. If its speed changed as the earth moved, electrons moving inside a computer could have different velocities, depending on how the device was oriented to the planet, or on which direction they were traveling inside the machine. Michelson–Morley's error factor of one in ten thousand might not seem like much. But it is a hundred times greater than the modulation in most microwave-based communications systems. Analog TV is unusable without signal-to-noise ratios of 50 decibels—one hundred thousand to one. Shifting light speed would skew the nanometer channels of wavelength division multiplexing many colors of light down a single fiber thread. It would disrupt the machines used to scribe patterns smaller than the wavelength of ultraviolet light on microchips. It would confound the global positioning satellites that guide your Never Lost Hertz rental car. And it would not help those cruise missiles racing across enemy skies.

By the beginning of the twentieth century, Michelson and Morley had plunged conventional thinking about light speed into a state of prolonged crisis. Maxwell had calculated it as a constant half a century earlier. But that idea did not conform to the consensus (including Maxwell's own belief) that the earth moved through a fixed ether, and that light speed therefore *had to be* variable, dependent on the movement of its source. Then there was Newton's ancient principle of relativity—that it is impossible to measure an object's velocity except by reference to some other object. Could light be the absolute exception?

For two decades, Albert Einstein ruminated on the perplexities. He dreamed of racing a wave of light: Seen by an outside observer, it would move at the usual pace of 300 million meters per second; seen by him, it would be at rest. But the very idea of light at rest struck Einstein as an impossible paradox. Maxwell had defined light by its frequency, which multiplied by its wavelength always equaled the mystical 300 million meters per second. How could a frequency, which is pure motion, be at rest?

As Einstein pondered the problem, he gradually realized that if Maxwell's view of the speed of light as a constant was to be upheld, Newton's classical mechanics, based on a fixed time-space grid, would have to give way. If light speed is to be an absolute, Einstein concluded, time and space had to be able to change. This insight launched the most fundamental revolution in the history of science. Einstein took the very gauges and indices by which physicists had always grasped the world, and substituted elastic, curved, ever-changing parameters. Later, in his theory of general relativity, light itself would bend to an increasingly elastic grid. Time and space distended into a curvacious, four-dimensional synthesis.

Of course, none of that was apparent, any more than the earth's curve is evident on your front lawn. Millennia passed before mankind discovered the spherical shape of the globe. More centuries passed before Einstein conceived the elastic universe. It is only in grasping the enormity of the cosmos, in which the earth is a marvelous but infinitesimal mote, that light-years and gravitational bends start to mean anything significant.

Similarly, technologists did not have to face the reality of the speed of light as an absolute limit of their technologies until, like Einstein riding his mind beam, megahertz computers and terahertz fiber-optic networks began to crash into the light speed wall.

Well into the past decade, the speed of light was an incidental factor. Most electronic devices operated too slowly for light's speed inside them to constitute a limiting factor. But as engineers wring out every inefficiency and elasticity, that blissful ignorance has changed. Reduced-instruction-set computing has multiplied the number of instructions performed per second into the billions and the time per instruction into the nanoseconds. Here the speed of light becomes consequential even across a tiny microchip. In fiber optics, gone are most of the time-hungry electronic line amplifiers and converters, replaced by all-optical amplifiers whose only real restraint is light's own speed. In thousands of satellites in geosynchronous orbits 23,600 miles above the earth, engineers have pared the protocols of voice traffic until the switching delays dwindled to a few hundred milliseconds. All these heroics had the effect of eliminating delay (or latency) from information systems.

Finally, the engineers have reached the end of the line. They have exhausted the tricks of their trade. Now they starkly confront, like Einstein, the residual barrier of absolute light. Like the physicists before them, they collide with it. And everything shatters—computer architectures, network topologies, satellite systems, software conventions—the entire time-space grid of the information economy.

The most immediate collision is in satellite technology. Light speed alone requires at least 250 milliseconds for a round trip to the Clarke belt, the geosynchronous orbit 23,600 miles away defined by Arthur C. Clarke. Two hundred and fifty milliseconds is a tolerable delay in voice communications but it is untenable for interacting with data. In 250 milliseconds, a modern computer would transmit the equivalent of a thousand large books before it would receive back any acknowledgment, error correction, or other feedback. A gigabyte of data would be afloat, stored en route, buffered in air, wires, or other media, and impervious to the wishes and keystrokes of the user.

The response has been a new industry of low-earth-orbit satellites. Loral–Qualcomm's Globalstar, Alcatel's SkyBridge, and Gates–McCaw's Teledesic are collectively launching thousands of devices into tracks twenty-five to fifty times closer to the Earth than the Clarke belt. One fiftieth of 250 milliseconds—five milliseconds—is a level of delay as good as the latencies of local fiber-optic systems already operating between New York and Washington.

On the ground, network delays stem chiefly from the distances between the Internet routers that take in a particular message packet, read its address, and forward it to a particular destination by the best available path. In the fastest routers of 2000, this process took ten microseconds (millionths of a second); soon it will be a tenth that. The problem is that across the Internet, an average message flows through seventeen routers, some messages hit as many as forty routers, and most web pages contain diverse elements that collectively take as many as twenty back-and-forth trips to download. Many of the routers are thousands of miles, or tens of milliseconds apart. The delay budget for voice communications is 150 milliseconds (more than that and you step on each other's conversational toes). The dream of ubiquitous voice and video over the Internet depends on reducing delays across the network. Worldcom's UUNet can guarantee 80-millisecond latency to its business customers, but the average Internet packet takes over 160 milliseconds in the U.S. and twice as much overseas. The net needs a new topology.

Similarly on microchips, computer operations entail accounting carefully for the speed of light. In a medium, the light-speed limit varies in proportion to the medium's index of refraction or dielectric constant. In metal wires, this translates to nine inches—twenty-three centimeters—per nanosecond, against about one foot per nanosecond in a vacuum. Such calculations deeply affect the newest microprocessors. A 600-megahertz Pentium II, for example, may execute one computer instruction in a nanosecond, the time needed for a signal to move nine inches on a metal wire.

Nine inches per billionth of a second might seem ample on a microchip with features measured in the microns (millionth of a meter). But a leading edge chip today bears as much as *seven miles* of wires. Even the Pentium II in your desktop PC has some 400 meters of wire on chip, about a quarter mile.

Synchronizing the movement of signals across the chip is the pulse of a piezoelectric crystal clock. An 800-megahertz chip—a high-end Pentium III, for instance—has a clock pulsing 800 million times a second. To coordinate all the functions of the processor, the clock pulse must reach all portions of the chip within the time that it takes for a transistor switch to flip from on to off. But this is the last generation of processors in which the pulse can actually reach across the entire chip within that timespan. In the .10-micron devices now emerging from laboratories, the clock pulse will be able to reach only 16 percent of the chip.

Hitting the light-speed barrier, the chip's architecture will necessarily fragment into separate modules and asynchronous structures. We might term these processors now under development time/space *mollusks*—Einstein's word for entities in a relativistic world—with their size limited by the systole of a light pulse. Setting the size of the integrated circuit module will be a measure tantamount in the microcosm to light-years in the cosmos.

Moving off the silicon surface, the tiny electronic charges boost themselves down the metal pins of the package, which are the legs of the millipedelike chip. Then these signals run out onto the expanse of the backplane buses and motherboards of a computer. That too takes time. Once you leave the chip surface and move to the pins, the problem begins to get more acute. The so-called Rent's rule states that pins multiply by the square root of the number of transistors. In other words, while the number of transistors rises from seven million on a Pentium II to a hundred million on a new generation processor—a factor of sixteen—the number of pins ekes up by a factor of four. Some of these pins are for power and ground connections, not data. Nonetheless Rent's rule makes it progressively harder to get data off the chip where you can use it. William Dally of MIT estimates that by 2010, the number of transistors will have risen one thousandfold and the number of data pins tenfold since the mid-1990s. With millions of times more transistors than links to the outside world, the chip faces a light-speed crisis that requires a radical change in the time-space relations of processors and memories.

Until recently, the effective speed of Intel microprocessors was measured by the clock rate, which mostly depended on how closely together its chief technical officer Gerry Parker could pack transistors on a chip. For the last several years, however, the effective speed of the microprocessor depends not on the clock rate but on the time it takes to fetch an instruction from memory. Richard Sites of Compaq subsidiary Digital Equipment Corporation estimates that the newest Pentiums and Alphas spend 80 percent of their time in wait states, marking time while the signals dawdle down the pins and across the bus to retrieve the data needed to perform instructions.

In sum, this problem is latency, the growing gap between processor speeds and memory access times. You can gun up clock rates and expand bus sizes and increase the bandwidth and capacity of memories by throwing money at the problem. A few billion dollars per year per fab will do it. But latency—the delay between the issuing of an instruction and the retrieval of needed data from memory—is determined by the time it takes for that round trip from the processor down the pins to the needed memory address, and that delay is set by the speed of light. Money won't change it; you cannot bribe God.

Computer performance is now governed far more by memory speeds than by processing speeds. The only way around it is to stay on the chip as much as possible. This means single-chip systems and portends deep changes in the relationship between memory and processor. Today memories comprise perhaps 99 percent of the computer's silicon area, and processors capture 75 percent of the profits. Thus processor manufacturers have been gaining some 400 times more profit per transistor than memory producers have. This financial gap will be closed in the future with single-chip systems. In essence, it will become more attractive to put processors on memories than to put memories on processors.

Moving a message across a network is similar to moving it across a computer bus; both hit the light-speed wall. While the new single-chip systems will integrate an entire computer on a single silicon substrate, the new networks will inscribe computers on the mostly silicon substrates of continents and sea beds. If the speed of light gets you even on the tiny expanse of a single chip, it dominates every calculation of computer networks. Here the relevant metric is how many meters a signal can travel in a microsecond or how many microseconds it takes to travel a kilometer (about five).

As bandwidth increases, an ever larger share of round-trip time is light-speed latency. Going between Boston and Albany, for example, bandwidth will hasten your trip only to the point that all the traffic is accommodated. Beyond this point, you cannot accelerate the trip down the Massachusetts Turnpike merely by adding new lanes to the road. Regardless of the bandwidth of the highway, physics imposes the ultimate speed limit: the velocity of light.

The only way to combat light speed is by moving closer to the data or moving the data closer to you. On networks, light speed enforces webs that use the same kind of caching systems previously employed on chips. Emerging is a "storewidth" paradigm that focuses on copying, mirroring, replicating, and storing web pages as close as possible to the final destination. Storewidth is a metric that combines storage and latency—it's not only how large and fast the disk drives are on the net but how close they are to the user and how quickly they can respond to a query or request. This change exalts directories and search engines into the crucial facilities on the Internet.

The law of the microcosm returns with a vengeance in the telecosm. Far from bringing a concentration of computing, the telecosm is shattering the integrated and centralized networking dreams of telecommunications companies into millions of subnetworks and Internet nodes, interconnected through the web's protocol. Based on one protocol and one optical technology, these networks in a sense will be more integrated, more unified, than any networks of the past. A planetary utility, they will be unitary and centralized on our single globe. But this planetary utility will release a fabulous diversity on its edges. The systole of compression will launch a diastole of decentralized intelligence around the world.

The new gigahertz computers and gigabit and terabit networks of the twenty-first century will comprise a fundamentally different architecture based on a different calculus of abundance and scarcity. Whether in a single computer or in a global network, the nemesis will be delay, or what computer scientists call latency. It refers to the loss of time in delivering the first bit, whether tapping a memory or crossing a network. In the new era, latency will become king.

As Leonard Kleinrock warned the industry back in 1992, in the new regime, communications speed "is of no use at all in speeding up the transmission of a file; it is the latency of the channel that dominates the time to deliver the file. The speed of light is the fundamental limitation for file transfer in this regime! And the speed of light is a constant of nature which we have not yet been able to change."

Engineers have a slogan: Bandwidth problems are solved with hardware, latency problems with software. With software advancing at only about one tenth the speed of hardware, this shift of emphasis would seem to portend a drastic retardation of the pace of progress in the industry.

The answer will be a new paradigm. At the core of the net, software will harden. An all-optical fibersphere, mostly devoid of software intelligence, will prevail. On the edges of the network, where data are delivered to servers and personal computers, hardware will soften. Rather than hardwired telephones and television sets, or even systems like the Pentium dominated by fixed instruction sets, new software based on adaptable processors will prevail, using new kinds of software components delivered just in time across the net. Whether you want to play a game, read a "newspaper," or conduct a video teleconference, your device will be adapted on the spot for the purpose at hand.

On all sides the information economy is bounded by the characteristics of light. Just as light constitutes the critical limit of information technology, light also supplies the redemptive abundance of the new hardware era. Measured in gigabits per second per kilometer without regeneration, bandwidth will increase over the next decade at least three times the pace of Moore's law. New technologies of sand, glass, and air will form a web with a total carrying power, from household to global crossing, at least a million times larger than the network of today. By the end of this tidal ascent, the global information economy—from the internal architectures of chips to the topologies of networks—will assume an entirely new shape, responding to the entirely new matrix of scarcity and abundance.

The remedy for the limits of the speed of light in communications is not more processing but less, not network software but network hardware, not intelligent switches but dumb pipes of boundless bandwidth. To transcend the constraints of circuits integrated on silicon substrates and spread across mostly silicon continents, the industry must consummate its destiny of overthrowing matter. It must replace current systems based on the heavy lifting of massy electrons by the trillions per bit with the massless, passive routing of photons, weaving the limits of light speed into an abundance of spectrum. We must move into the fibersphere where networks of photons serve a penumbra of software on the edges: networks of light linking cathedrals of mind.

TELECOSMTELECOSMTELECOSMTELECOSMTELECOSMTELECOSMTELECOSMTELE-
COSMTELECOSMTELECOSMTELECOSMTELECOSMTELECOSMTELECOSMTELE-
COSMTELECOSMTELECOSMTELECOSMTELECOSMTELECOSMTELECOSMTELECOS
MTELECOSMTELECOSMTELECOSMTELECOSMTELECOSMTELECOSMTELECOSMT-
ELECOSMTELECOSMTELECOSMTELECOSMTELECOSMTELECOSMTELECOSMTELE-
COSMTELECOSMTELECOSMTELECOSMTELECOSMTELECOSMTELECOSMTELECOS

PART TWO

MTELECOSMTELECOSMTELECOSMTELECOSMTELECOSMTELECOSMTELECOSMT-
ELECOSMTELECOSMTELECOSMTELECOSMTELECOSMTELECOSMTELECOSMTELE-

THE NEW PARADIGM

COSMTELECOSMTELECOSMTELECOSMTELECOSMTELECOSMTELECOSM
TELECOSMTELECOSMTELECOSMTELECOSMTELECOSMTELECOSMTELECOSMTELE-
COSMTELECOSMTELECOSMTELECOSMTELECOSMTELECOSMTELECOSMTELE-
COSMTELECOSMTELECOSMTELECOSMTELECOSMTELECOSMTELECOSMTELECOS
MTELECOSMTELECOSMTELECOSMTELECOSMTELECOSMTELECOSMTELECOSMT-
ELECOSMTELECOSMTELECOSMTELECOSMTELECOSMTELECOSMTELECOSMTELE-
COSMTELECOSMTELECOSMTELECOSMTELECOSMTELECOSMTELECOSMTELECOS
MTELECOSMTELECOSMTELECOSMTELECOSMTELECOSMTELECOSMTELECOSMT-
ELECOSMTELECOSMTELECOSMTELECOSMTELECOSMTELECOSMTELECOSMTELE-
COSMTELECOSMTELECOSMTELECOSMTELECOSMTELECOSMTELECOSM
TELECOSMTELECOSMTELECOSMTELECOSMTELECOSMTELECOSMTELECOSMTELE-
COSMTELECOSMTELECOSMTELECOSMTELECOSMTELECOSMTELECOSMTELE-
COSMTELECOSMTELECOSMTELECOSMTELECOSMTELECOSMTELECOSMTELECOS
MTELECOSMTELECOSMTELECOSMTELECOSMTELECOSMTELECOSMTELECOSMT-

Chapter 5

The Road to the Fibersphere

Tracking down the origins of the telecosm, an investigator might be surprised to find himself on the streets of Greenville, South Carolina, in the mid-1930s. Perhaps it was something in the air, or in the light. Perhaps it was a local piano teacher named Bernard Einstein, whose famed photo-physicist uncle was a celebrated visitor. Whatever it was, this sleepy southern city was pregnant with photonic portents.

One of Greenville's spectrally obsessed sons, Charles Townes, would go on to invent the laser, which won him a Nobel prize. Another, Irving Goldsmith, became DuPont's chief scientist, famed for his own optical inventions.

The last of Greenville's spectronic homeboys—all students, incidentally, of a high school physics teacher with a touch of magic and a telecosmic name, Belle Free—is a lanky, courtly man named Will Hicks. His fame came as coinventor of single-mode optical fiber—the gossamer strands of glass that are the world's supreme carriers of information. But he has another role to play as well: father of the telecosmic vision.

Hicks was born and raised in Greenville, a city riven by a gully holding a black ghetto where his landlord father owned thirty-five houses for rent. By age fourteen, Will was already helping his dad as a carpenter, plumber, and electrician and managing five employees. The young Hicks could solve scientific riddles in his head and had what he fatefully recalls as "an intuitive sense of the calculus." Through his piano teacher he met the hoary paladin of relativity on no less than three occasions. But even in Belle Free's clearly capable hands, classical optics—the study of rays and beams and lenses—put him to sleep. He also claims to have slept through optics classes at Furman in Greenville, where he ended up with a degree in physics, and in the wartime navy, which sent him to classes at Harvard and MIT to learn about radar. Finally, relegated to the far fringes of the Manhattan Project, at a secret facility in West Virginia, Hicks woke up to the pleasures of silica. In a search for better lab containers, he spent pleasant hours on end in the machine shop, melting and blowing glass.

After the war, Hicks took his young family out to Berkeley, where the famed Radiation Lab essayed to teach Hicks high-energy proton scattering, meson interactions, and quantum field theory—the great scientific frontiers of the day. In the late 1940s the Radiation Lab was a Parnassus, with Victor Weiskopf, William O. Lawrence, Edwin MacMillan, Robert Oppenheimer, and W.K.H. Panofsky walking the halls. Awed, Hicks was drawn to field theory's abstruse challenges. As part of his course work, he even contrived some numbers so arresting that they made their way into a speech by the great Oppenheimer himself. There was only one problem: Hicks's numbers, "intuitively calculated," turned out to be horribly wrong. Amid the embarrassment, Hicks fled without a doctorate. So he would not devote his life to the baffling enigmas of the quantum. But he could blow glass, and had a deep, practical sense of the behavior of optical materials. For some purposes, these skills turned out to be worth more than a barn full of Schroedinger's cats (and equations).

By the early 1950s, Hicks was back across the country again, at American Optical in Southbridge, Massachusetts, making bundled glass fibers for the CIA to use with spy cameras. "From day one," he discovered, everyone was treating the problem in classical optic terms. Ray theory predicted that a beam of light could be channeled down the fiber—provided you could wrap the glass with perfectly reflective cladding, which no one had managed to do. But from day one, Hicks ignored ray theory, easy enough for him to do since he hardly knew it. Instead, he confidently applied Maxwell's nearly century-old electromagnetic radiation theory, with its refraction indices and dielectric constants. Thus he could predict that if fibers were drawn thin enough, light of just the right frequency passing through would resonate in the glass. Instead of bouncing crazily off theoretically mirrored walls—in the process blurring into near uselessness—photon after photon would flash through the fiber in a "single mode" like freight cars hurtling down a track. This concept of single-mode fiber ended the seemingly insurmountable problem of transmitting light through more than a few inches of glass fiber.

The greatest of journeys begins with one small step. But no one at the time imagined that the capability of sending light a foot down a waveguide could lead to transmitting it, twenty-five years later, five thousand miles under the oceans. Hicks's insight led him in 1957 to copublish a paper declaring the impregnable superiority of so-called single-mode fiber—what was then a useful material for scoping a stomach, but today is the key material of the telecosm. With a core just five times as wide as the wavelength of the infrared light (which was less than two microns), a single fiber now has a theoretical capacity as high as 25,000 gigabits per second of information. This is enough to carry more than three million large books a second (a million bytes—a megabyte—is roughly equivalent to a hundred thousand words of text).

Were it not for another of the muddled calculations that were Hicks's trademark—this one relating to dispersion of light in the fiber—he might have realized the potential of fiber for long-distance communications. Instead he started a company of his own, Mosaic, to transmit images through a few feet of fiber, for endoscopes to survey internal organs or for fiber devices to encrypt pictures for the CIA. Hicks still failed to recognize the significance of his inventions, still failed to see more than a few feet down the road to the fibersphere.

A larger vision entailed a sense of the quantum theory that Hicks had fumbled at Berkeley. Both the telecosm and the microcosm spring from the quantum, with its paradoxical synthesis of definite and infinite, particles and waves, compression and expansion.

• • •

"To see a world in a grain of sand . . ." is a vision of William Blake, framed in a sampler on the wall by my bed. Blake's words offer a poetic summation of the microcosm: worlds inscribed on grains of sand, the silicon chip. Wrought of sand, oxygen, and aluminum, the three most common substances in the Earth's crust, the microprocessor distills ideas as complex as a street map of America onto a sliver of silicon the size of a thumbnail. This gift of the quantum is a miracle of compression. The gift of the telecosm is a miracle of expansion: grains of sand spun into crystalline fibers and woven into worldwide webs.

Throughout the realms of information technology we find reflections of the quantum duality of particles and waves, compression and expansion. Electronic transistors use electrons to control, amplify, or switch electrons. Electrons have mass and charge and impact other electrons. But waves differ radically. Because moving photons do not affect one another on contact, they cannot readily be used to control, amplify, or switch one another. Compared to electrons, moreover, photons can be huge: At 1,550 or 1,300 nanometers, infrared wavelengths are larger than a micron across. They resist the miniaturization of the microcosm.

But the real photonic showstopper is storage. Photons can scarcely be stored or buffered at all. For computing, photons are far inferior to electrons. With single-electron electronics now in view, electrons will keep their advantage. For the foreseeable future, fast computers and network packet switches will be made with electrons.

What are crippling flaws for computing, however, are huge assets for communicating. Because moving photons do not collide with one another or respond to electronic charges, they are inherently a two-way medium. They are immune to lightning strikes, electromagnetic pulses, or electrical power surges that destroy electronic equipment. Virtually noiseless and massless pulses of radiation, they move as fast and silently as light. Listening to the technology, one finds a natural division of labor: electronics for computing, photonics for communications.

The fulfillment of the photonic dream, though, faces persistent obstacles. One is chromatic dispersion, the tendency of different "colors" of light—different wavelengths and frequencies—to move at different speeds down a fiber. As Maxwell showed, refractive indices affect different colors differently. Since even a pure laser signal covers a significant band of spectrum, the signals would tend to mush as the light passed down the thread.

Modal dispersion, on the other hand, refers to the tendency of different waves of the same frequency to take different paths down the fiber—some zooming along the core, some bouncing along the cladding. Different paths mean different elapsed times, even when the pulses travel the same speed. Again the result is a blurring of the signal as it passes down the fiber.

The first to point to solutions for many of these problems was Will Hicks's collaborator at American Optical, Elias Snitzer, coauthor of the historic paper on single-mode fiber. Another onetime Communist like Gould, who also had to struggle to gain government clearance to inspect the fruits of his own inventions, Snitzer at least had a physics Ph.D. He proved his mettle as an inventor and scholar through a creative span of fifty years that continues today at the Optical Center of Rutgers University in New Brunswick, New Jersey.

Amid the tumult of scientific claims and Nobel laurels in the then flourishing world of gas lasers, Snitzer's experiments at American Optical were so far ahead of his time as to be nearly invisible to his fellows. Snitzer's focus was not on gas or rubies, but glass. He was seeking to amplify light using silica, impregnated with rare-earth elements, such as erbium and praseodymium. No one took him seriously, since no one had any notion of a way to use glass with lasers in communications, and rare-earth elements just complicated the already murky domains of solid-state. Gordon Gould's idea of using lasers to spark nuclear fusion was seen as a more likely application.

Also during the early 1960s, Snitzer and Hicks experimented fruitfully with methods of cladding fiber with other glasses with lower refractive indices, rather than with the laquers, compounds, and plastics that were popular at the time. These experiments foreshadowed today's fiber-optic threads similarly cladded with glass for "total internal reflection," so no light is lost in transmission down the fiber.

Fiber-optics communications technology, however, would not emerge for several more years. A key step was the 1966 finding of Charles Kao and George A. Hockham at ITT in Britain that the major obstacle to communicating through glass was not the material itself but impurities in it. With the impurities removed, they calculated that fiber could theoretically bear light as far as a kilometer while retaining a readable 1 percent of the signal. One percent sounds small, but if it could be achieved, it would represent a landmark in the history of fiber optics.

Kao emerges as one of the heroic innovators in the history of technology. An electrical engineer, he had the intuitive sense to ask the right question—not "What is the attenuation in the conventional glass existing in the world?" but "What is the attenuation of theoretically pure glass?" Plugging in the evidence for the losses in pure glass, he showed that an intractable form of pure glass called fused silica would have virtually no attenuation at all. In the mind of one visionary at least, the fibersphere had been invented. That was enough. Campaigning militantly and tenaciously over the next five years for the use of fiber optics in communications, he finally found the crucial converts who could make the theory real.

The key was persuading Robert D. Maurer, a physicist at Corning Glass with a team pursuing the possibilities of fiber communications. Like most investigators of the 1960s, Corning researchers had focused on compound glasses that are more robust and easier to bend and simpler to manipulate chemically for the desired refractive index and other properties. But Kao's findings inspired them to switch to pure silica despite its brittleness, high melting point, and other obstacles to mass manufacturing and installation. One obstacle was that cladding materials available for mass-produced pure glass wires (in contrast to Snitzer laboratory devices) had a higher refractive index than the core, meaning that the light would leak rapidly away.

Maurer and his associates at Corning—Donald Keck and Peter Schultz—had to develop an entirely new technology for glass fiber creation. Turning common sense upside down—getting dressed from the overcoat in—they began with the cladding, creating a hollow tube, open in the middle. Then they infused the tube with silica, doped with other elements that increased its refractive index above the index of the cladding.

The first fibers created by this technique showed prohibitive signal losses over any significant distance. But by 1970 Maurer and his team managed to exceed the Kao one-percent standard for effective transmission (20 decibels lost per kilometer) in the 630-nanometer band of orange light that could be produced by a helium neon laser. Corning soon introduced a next generation of fiber that lowered the attenuation to a level of under one decibel per kilometer. Moreover, a Corning engineer named Frank Hyde figured out an efficient way to manufacture fused silica in volume.

Still, fiber optics remained at an impasse, restricted to relatively short-distance applications. AT&T favored satellites and millimeter waveguide pipes for its next generation of long-distance links. MCI was building networks of microwave towers that communicated by line of sight. In fiber, meanwhile, the two kinds of interference inherent in the glass—chromatic and modal dispersion—greatly reduced the distance that the signals could travel without expensive electronic repeaters that destroyed most of the economic incentive to deploy the technology. No one was applying Hicks's and Snitzer's single-mode waveguide concept or Snitzer's in-fiber amplification using rare-earth elements.

In the late 1970s, at a time when a solution seemed most remote, Maurer's team at Corning made a further breakthrough. Pushed by the findings of Snitzer and Hicks—and a visit from Hicks during which he memorably spliced two single-mode strands in his own hands—the Corning team fixed on single-mode fiber. They proved the Snitzer–Hicks proposition that by reducing the size of the fiber to under ten microns they could restrict the waves to a single mode or path down the fiber.

With single-mode fiber there would be no modal dispersion. At the same time, Maurer and his colleagues discovered that new semiconductor lasers—small enough to couple to the tiny ten-micron aperture of the single-mode fiber of the day—operated at an infrared wavelength (1,310 nanometers) that intrinsically reduced chromatic dispersion to zero. In 1977, engineers at MIT's Lincoln Laboratories, led by J. J. Hsieh, had invented semiconductor lasers of indium gallium arsenide phosphide with lifetimes of 1,500 hours at the very 1,310-nanometer wavelength that eliminated chromatic dispersion. At this wavelength, an offsetting compression effect nullifies the tendency of the signal to spread out as it moves down the fiber.

Later a similar invention produced semiconductor lasers operating at the 1,550-nanometer band. Though less effective than 1,310 nanometers for limiting chromatic dispersion, the new laser did cause attenuation (signal loss) to drop to a minimum. There were now two highly effective infrared semiconductor lasers: one 1,310 nanometer, ideal for short-distance communication, and the other 1,550 nanometers, with low attenuation ideal for long-distance communication. Both optimal wavelength channels turned out to be in the very infrared band that happened to define the only kind of semiconductor laser that it was possible to make. The telecosm stems from such miraculous coincidences, comparable in importance only to the discovery that silicon could be both electrically insulated and chemically protected by its own oxide—a finding that had been similarly crucial to the development of the microcosm.

By the time all these technologies came together in the late 1970s and 1980s, few people remembered that Elias Snitzer had been the prophet and inventor of most of them. But Snitzer's most important contribution was not to manifest itself until the 1990s. It originated in the summer of 1960 when he began to experiment with glasses doped with elements known to fluoresce (i.e., convert energy into light) at visible and infrared wavelengths. Fluorescence is a preliminary stage in lasing. To obtain the desired infrared emission, Snitzer used the rare-earth elements neodymium, praseodymium, and erbium. Neodymium turned out to fluoresce at 800 nanometers, praseodymium at 1,320 nanometers, and erbium at 1,550 nanometers—all in the infrared bands that fit the communications sweet spots of optical fiber.

In the late 1990s, these researches achieved a stunning new vindication. Snitzer had proposed an erbium-doped glass amplifier of photonic signals that could obviate their conversion to electronic form for regeneration. (Otherwise, fiber optics is limited to 35-kilometer distances and you can form a network only if you contantly shift the signal back from optics to electronics. This, in fact, is what the first fiber-optic networks did.) Thus along with his role in cladding and single-mode fiber, Snitzer was also a father of wavelength division multiplexing (WDM). These systems, now enhancing a millionfold the potential capacity of fiber-optics networks, send many different "colors" of infrared light down a fiber at once, each wavelength bearing a different stream of messages or bits. No pauses for expensive electronic pit stops.

Operating in the unique 1,550-nanometer low-loss band, the broadband amplifiers pioneered by Snitzer over a period of some thirty years make wavelength-separating and electronic-converting and processing unnecessary. When the dying photons in the line hit the complex erbium ions in the amplifier loop and are excited by a semiconductor "pump" laser, the ions automatically emit floods of new photons identical to the original signal. Although an electrical source powers the pump laser, the erbium-doped amplifier is essentially an all-optical device, requiring no conversion or processing of individual signals. This feat enables any given mixture of signals to travel from origin to destination entirely on wings of light—Look, Ma Bell, no electronics—under the seas and across the continents of the globe.

At the time of Snitzer's early rare-earth amplifiers, however, fiber optics was still in its infancy and no one grasped the importance of the invention. It was not until 1985 that David Payne and Simon Poole at British Telecom Laboratories and Southhampton University first began reporting experiments with neodymium-doped glass. It was not until 1988 that they first proposed an erbium-doped amplifier and claimed that they had invented it—apparently unaware of Snitzer's prior work. Soon afterward Emmanuel Desurvire and Randy Giles of Bell Labs built workable rare-earth devices.

Now an elderly professor emeritus at the Rutgers Optical Center near Bell Labs in New Jersey, Snitzer and his students are still in the forefront of optical investigations. In 1991, Snitzer finally received the coveted Charles H. Townes award for outstanding work in quantum electronics. Along with his 120 patents and 70 archival papers—most on various forms of doped glass lasers and waveguides—the Townes tribute exalted Snitzer's creation of the first glass laser and the invention and first operation of erbium-doped glass lasers.

Snitzer and his students are now working on tellurite fiber lasers for future amplifiers with double the bandwidth of the existing erbium-doped devices. The wider the bandwidth of the amplifiers the larger the number of discrete wavelengths (and thus separate bitstreams) that can be sent down a single-fiber thread.

Expressing pride both in his Republican son-in-law and in his early resistance to the House UnAmerican Activities Committee, Snitzer is now an honored academic figure. He reveals no sign of envy toward the Nobel prizes of others, or resentment toward the fortunes made by people using his inventions or toward the claims of others to precedence in areas where he was decades ahead. The single-mode fiber and the cladding invented by Snitzer and Hicks, and the erbium-doped amplifiers pioneered by Snitzer, were the keys to the telecosm. But the telecosmic technologies would not be built by either.

The practical breakthroughs came instead out of the depths of IBM's huge Watson Laboratories, followed close behind by research centers around the globe: NTT's Yokosuka facility, British Telecom's Martlesham Heath, and IBM's own Rushlikon Labs in Zurich. Often built from off-the-shelf components, these systems were united by a powerful new idea, an infinitely supple rival to the existing labyrinths of telecom: the all-optical network.

As an ideal, all-optical networking eschews complex electronic switches, creating a fibersphere as neutral and passive as the atmosphere. Though they may be thousands of miles apart, computers will communicate across it the same way radios and TVs "connect" through the air—directly, by unruffled streams of wave-borne information. Just as Bob Noyce's integrated circuit has ultimately led to entire computer systems etched on a single sliver of silicon, all-optical networks will enable an entire communications system to be built on a seamless seine of silica—fiber-optic glass.

The progenitor of the all-optical network—he coined the term, built the first fully functional system, and wrote the definitive book about it—is Paul E. Green Jr., then of IBM's Watson Labs, later at Tellabs Corp., the aggressive Florida-based firm that produces telecom capital gear. A shambling southerner with an engaging drawl and a hard edge of technical savvy and ambition, Green comes from the same generation and region as Will Hicks. But Green is not a laser physicist, an optical engineer, or even a telecommunications expert. At IBM, his work ranged from overseeing speech-recognition projects to shaping corporate strategy. He wore the IBMers' white shirt and stumbled into telecosmic history almost by accident.

As a computer man, Green relished the contrast between the onrushing efficiencies of Moore's law and the phone system's torpor. In the quarter century that started with the introduction of integrated circuits around 1960, computer powers rose a millionfold, network capacities merely by a factor of a thousand. Not until the late 1980s did most long-distance data networks surpass the bit rates achieved by the Pentagon's 1960s-vintage ARPAnet, the Internet precursor that ran at fifty kilobits per second.

Green's optical epiphany came during a 1985 technology debate at Cornell which pitted him against Robert Lucky, then of Bell Labs and now of Bellcore. The inventor of adaptive equalization for long-distance telephony and the industry's leading prose stylist, with five books to his credit, Lucky was a phone man to his fingertips. When Green pointed to the telcos' increasing dependence on the high-flying computer industry, Lucky protested how unjust it was to compare the two fields: the microcosmic computer world, jamming ever more tiny wires and switches on single slivers of silicon; and the macroworld of laying wires across oceans and continents. "How would it be possible," Lucky asked, "to make such a large-scale system inexpensive?" The problem, he suggested, was inherent in the very nature of long-distance communications. The network itself was the bottleneck.

Green, the computer man, knew too little about the telecom world's byzantine details to grasp the limits that seemed so obvious to Lucky. He believed communications could achieve miracles comparable to computing's integrated circuit—if only you eliminated what he saw as the real problem: the pyramid's utter dependence on huge, complex, deceptively "fast" switches—the electronic machines that make the connections for users across the network according to the digits in phone numbers.

Overthrowing the switching model was (and largely still is) telco heresy. Telephone doctrine was historically rooted in the mathematic visions and theories of Claude Shannon, reclusive genius of Bell Labs and MIT. Shannon's "information theory" predicts the absolute carrying capacity of any communications channel. His ingenious formulae for relating signal-to-noise ratios to the capacity of a carrier were a key tool for opening up the postwar world of bandwidth scarcity. Information theory shows engineers precisely how to use switching to make up for inadequate bandwidth, by relieving a network of the need to broadcast all signals to every destination. Each message or stream of bits goes exactly where it's intended, and nowhere else. The highest of all telco virtues, bandwidth is conserved.

Benefiting from the exponential upwelling of computing power, Bell System engineers created enough transistor-driven intelligence to surmount every limit of copper wire. The drive continues today, with such technologies as integrated services digital network (ISDN) and digital subscriber line (DSL)—triumphs of microprocessors and clever algorithms over the confines of twisted pair wiring. Their core competence, the Bells never forget to stress, is switching. Their vision is to make those switches the entering wedge for ever more elaborate information services—a cornucopia of exotic new technologies and features, climaxing with Yellow Pages that exfoliate into a gigantic global mall of full-motion video, where your fingers walk (or your voice commands echo) from Harrods to Yahoo, from Amazon to Jardines, from Dell to Akihabara, from Schwab to eBay.

There is just one problem: in a high-bandwidth world of optical communications, that strategy is a telecosmic dead end. Fiber offers error rates *ten billion times lower* and bit rates *ten billion times higher* than Ma Bell's copper wires. The most intelligent central switches ever conceived, let alone built, simply will not switch data fast enough to handle the colossal floods of an Internet whose traffic is rising by tenfold every year or so, a thousandfold every three to five years, a millionfold every six years to a decade. Against that tide, even the eighteen-month doublings of Moore's law don't stand a chance.

By contrast, when the problem of lagging computer speeds first arose in the world of microprocessors, computer architects responded with parallelism—multiple processors operating simultaneously on multiple streams of data. While each worked more slowly than the fastest serial processors, scores or hundreds of them working in parallel could far outperform the fastest serial machines. The same principle turned out to work far better for all-optical networks. By dividing conventional laser beams into ever more finely tuned wavelengths—individual colors, in effect, each carrying their own gigabits of data—optical engineers introduced massively parallel communications. Wave-division multiplexing, or WDM, is the equivalent of parallel processing, and has all of its advantages.

Even in their embryonic forms, all-optical networks offered more bandwidth for the buck than any other network technology. Paul Green's prototypes quickly became a commercial product, IBM 9729, operating at a gigabit a second over a total of twenty different wavelengths. Limited by its design as an LAN—local area network, workable for no more than fifty terminals—it found a niche connecting Wall Street trading floors with back offices across the Hudson River in New Jersey. Through the late 1990s, Rainbow was generating an estimated $110 million in annual revenues.

But Green's dream fell short. Unwilling to invest what it would take to make a real plunge into telecom—never an IBM strongpoint in any case—the company canceled further Rainbow development in 1997, and sold its optical networks division to Tellabs. Green and his brilliant protégé Rajiv Ramaswami—together with a team of six other researchers—converted their IBM offices in Hawthorne, New York, into a northern unit of the Florida-based Tellabs corporation.

Even as they were changing the sign on the door, however, Green's concept was exploding across the telecom universe. Over the next year, optical networking fuelled a minigalaxy of Wall Street supernovas—Ciena, Qwest, Level 3, and AT&T's own mighty offspring, Lucent. In late 1999, Ramaswami left to guide a company called Xros in Sunnyvale in the creation of all-optical micromirror "switches" which shifted the paths of wavelengths. This efflorescence of creativity had not been seen in industry since the birth of the microcosm, four decades earlier.

Indeed, by the millennium, fiber optics had utterly overthrown the *ancien* telecom *regime* of fast computers and slow wires. In a startling inversion powered by avalanching Internet traffic, it was suddenly computers—all those supremely intelligent switches topping the telecom pyramid—that blocked the way forward. It was this electronic bottleneck that Paul Green hoped to blow away with his all-optical networks—the huge gap between the capacity of fiber optics and the actual speed of quotidian communications.

"More bandwidth!" became a rueful joke by 2000. Would-be web surfers complained about the World Wide Wait. As more and more data sluiced across the globe by the week, the solution became more and more obvious: not to try to build faster and faster switches, but to get rid of switching as much as possible in an all-optical regime. The solution is the telecosm.

From Hicks's and Snitzer's photonic researches to the heroics of microelectronics, both sides of the quantum duality converge in a key lesson. In the telecosm, as in the microcosm, the less the space the more the room. All the supposed drawbacks of higher frequencies, smaller devices, and solid-state complexities give way to supremely powerful advantages. From the microwaves of satellite links to the infrared connectors of personal organizers or TV remote controls, the most powerful communications systems are essentially photonic or "spectronic," using quantum pulses of radiation to send immaterial signals. A consummation of the overthrow of matter, spectronics is the science of electromagnetic communications, wired and wireless, the physical foundation of the telecosm.

Will Hicks failed to build any of these systems that enabled the telecosm. But he can satisfy himself by serving as a father, if not of the telecosm itself, then at least of the telecosmic vision. In 1985, at age sixty-six, he read my book *Microcosm,* and wrote me a letter declaring that the impact of microprocessors would soon be dwarfed by the impact of fiber. Intrigued, I invited him to speak at the late Robert Fisher's conference on new technologies at Nichols College, a small school outside Worcester, Massachusetts, that specializes in entrepreneurial studies. Meeting the craggy, white-haired southerner, I had no idea what to expect. What I got was unique in my twenty-year career as a public speaker.

I had always taken for granted that in any assemblage of pundits, I would be the most cornucopian—the most hyperbolically assured that silicon could save the world. At Nichols College, I gave my usual tribute to the abundance of sand and glass and air, of microchips, fiber, and wireless, all exponentially advancing at the time. Hicks stood up and said that I grossly underplayed the potential of glass. I looked on in shock as this hard-headed entrepreneur, with six companies to his credit and scores of inventions in fiber optics, explained that bandwidth could be boundless. Not only I, he said, but the entire telecommunications industry were underestimating the potential by a factor of millions.

Every fiber, Hicks pointed out, could carry millions of times the pathetic trickles of bandwidth that the phone companies offered their customers. With appropriate trunking and branching across the new wavelength circuits of the fibersphere, every citizen could command a personal frequency phone number. Every phone could handle real-time image streams. Every call could be a video teleconference. The problem, Hicks said, was that the telecom industry's guardians saw fiber as a mere retrofit for copper wire, with superior bandwidth but its own set of problems. What they were missing was its revolutionary implication: the potential for whole-body replacement for switches and air and microwaves and computer displays and geosynchronous satellites—in other words, for the whole creaking, century-old edifice of global telecommunications.

Overthrown will be much of the received wisdom of the microcosm. As Robert Lucky observed in a forward to Paul Green's book: "Many of us have been conditioned to think that transmission is inherently expensive [and therefore] we should use switching and processing wherever possible to minimize transmission." This is the law of the microcosm. But, as Lucky goes on to speculate, "the limitless bandwidth of fiber optics changes these assumptions. Perhaps we should transmit signals thousands of miles to avoid even the simplest processing function." This is the law of the telecosm: Use bandwidth to simplify everything else.

"Too cheap to meter," originally said about the promise of nuclear energy, is already turning out to be true for something potentially even more valuable than mere electricity: the movement—in literally unfathomable quantities—of information, of ideas. In the end, the new regime will change the way we see reality itself: the ruling paradigms of thought.

A paradigm is an explanatory structure. The key source of paradigm theory is Thomas Kuhn's 1962 book, *The Structure of Scientific Revolutions.* Unfortunately, Kuhn chose to define *paradigms* as cyclical structures of knowledge which would predictably succumb to new structures. In short, a framework that explained nearly everything today would, with the next "revolution," become worthless.

A philosopher who saw history as a kind of academic debate, Kuhn failed to grasp how the cumulative advances of science manifest themselves in powerful and practical machines that in turn propel new advances. Thus, technological paradigms are neither artificial nor arbitrary: they are the governing force that underlies human progress. Apprehended by scientists, applied and tested by engineers, they reflect profound—and *permanent*—truths of the universe. The microcosm does not simply give way to the telecosm. The microcosm is a crucial foundation of the telecosm.

Every age defines itself by the resource it wastes. Our agrarian forefathers wasted human time. The Victorians wasted coal and iron, the twentieth century wasted electricity. Over the past decade, the world had to learn to waste transistors. Now it needs to learn how to waste bandwidth, and begin rebuilding the world yet again.

Buying a condo in the Cayman Islands from your living room in

Cairo. Converting a car stuck in a Philadelphia traffic jam into a physics class at Caltech or a bay window view of the North Sea into a three-dimensional Rio beachfront. Having dinner with your daughter at Dartmouth and your wife at home in Massachusetts while sitting in an office in Sausalito overlooking the San Francisco Bay. You've heard the bandwidth "fantasies" before. But there's nothing fantastic—it will happen.

Into the fibersphere, the journey continues.

Chapter 6

The Collapse of the Seven Layers

Only connect . . . but faster! The fibersphere's flash and the Internet's rush should not lead us to forget that what is happening is no more or less than a hypertrophy of networking. This is the prosaic matter of transferring information from one person to another. It is less than divine redemption, but more than most other activities it defines us as human. It also lies at the heart of all economics and finance and makes into the era's largest business opportunity the supplying of equipment for the global ganglia of the net.

Until the late 1970s, nearly everyone simply used machinery provided by Ma Bell and her minions around the globe. If you tried to attach any jot or tittle of alien gear to this hallowed system—such as a hush-a-phone horn on your handset—you might theoretically go to jail. AT&T and the world's PTTs (departments of posts, telegraphs, and telecommunications) held legal monopolies in their countries and plied armies of lawyers and lobbyists to defend them.

Today by contrast, some five thousand companies supply key components for the Internet. The communications scene has changed from a pyramid of Bell to a tower of Babel. No longer will one imperial company ensure that every handset, plug, codec, and central office switch can link seamlessly together from sea to shining sea.

Now the challenge is to interconnect all the myriad components of polyglot hardware and software. In this market, participants must interoperate with other networks, in different countries, on different kinds of equipment, using different forms of electrical power and communications media, in diverse languages and protocols, with varying speeds and urgencies.

The paradigm trade show of the telecosm, therefore, has long been InterOp, a semiannual gathering of the world's networking nerds that is devoted to interoperation. Now merged into Networld plus InterOp and bought by Softbank of Japan, it began in the late 1980s, when impressario Dan Lynch threw down a gantlet to the networking industry. In the age of the Internet, mere claims that your equipment can interoperate will not be enough; you must come to InterOp and prove it. Nearly anybody who is anybody in networking showed up at InterOp and made his pitch.

The pivotal part of InterOp happened before the show. In a sweltering cavernous room usually in Atlanta, San Francisco, or Las Vegas amid a centihertz hum of fans sprawled a pastel simcity of angular and curvacious boxes. They thronged with application-specific microchips and hopes for plug and play and a billion-dollar IPO, all interspersed with a green glow of telltale CRTs and Cisco envy and desire. Tangled cables and twisted-pair wire snaked among blinking lights and empty Coke cans and arrayed racks and bays of dry-humping and heaving routers, switches, stackable hubs, nodes, and bridges, as well as beer bottles and network interface cards and credit cards and duct tape and Tylenol. Recapping the chaos of machines was a menagerie of hundreds of nerds and boffins with beards and scopes and sniffers and denim shirts hanging out; Indians, Chinese, cable guys, paunchy folk with wild hair, spruce dudes with crewcuts, united only by an exotic common idiom of pings and rups and interrupts and traceroutes and seven-layer brains.

Bristling as it does with the acronyms and complexities needed even to think about connecting the congeries of computer and telecom devices, the networking world often seems inscrutable. But all is not, in fact, chaos. Sanctioned by the Geneva-based International Standards Organization, there is a voluntary framework for global networking, the open systems' interconnect model—rarely referred to other than by its bland acronym. OSI defines a seven-layer netplex, from the physical layer—actual wire or fiber—at the bottom to an "application" layer—

the human interface—on top. In between, it is all there, from Internet and hypertext transfer protocols, to LANs and WANs. If you want your device or system to work with anyone else's, OSI is where you begin.

As the netplex is convulsed with constant waves of innovation, the OSI model creaks and shudders but remains in place. Misleading in its specifics, it may err in describing a particular system or device on the floor of InterOp, or in the Internet. But it is universally respected as a heuristic tool, vital for impressing customers, reporters, employees, students, and for seeming savvy at trade show exhibits. It also offers a way to grasp and discuss the complex regimen entailed by any electronic communications system. So you should learn it. To help, I have contrived a misleading mnemonic: *phydlnets,* pronounced "fiddlenets."

Starting at the bottom with the physical layer (*phy*), the OSI layers move upward through datalink (*dl*), network (*ne*), transport (*t*), and session (*s*).They conclude with the actual presentation (*p*) and application (*a*). For a deceptively familiar example consider a phone call. Pick up the handset and listen for a dial tone (*physical layer*); dial up a number (every digit moves the call another *link* closer to the destination); listen for the ring (signifying a *network* connection and *transport* of signals). Getting someone on the line, you may be said to have completed the first four layers of the OSI stack. Then your hello begins a *session,* the choice of English defines the *presentation,* the conversation constitutes the *application* layer. The hangup ends the *session.* You may be said to have proceeded through seven layers or so, each dependent on the one before it in the stack. Thinking like this you are simulating a network engineer.

A materialist might suppose that the *physical layer* is all that there is. But in grasping how a network operates, it gets you almost nowhere. Incoherent and heterogeneous, it includes the backplane of your computer, the twisted-pair phone wires to your house, the fiber lines to the central office, the central office switch or Internet router, the fiber trunks across the continent or under the sea, or microwaves out across the ionosphere, and all the gear in between. Animated by millions of interconnected microchips containing trillions of transistors, the physical layer is fundamentally unfathomable. But above the physical network, rendering it functional and comprehensible, is a logical network, to which all the other layers belong.

A signal cannot pass a single stretch of the physical network without observing codes and protocols suitable to that physical link in the chain, whether a fiber line or a twisted-pair wire, a central switch, or a telephone private branch exchange (PBX). These media-specific codes and protocols, bit rates and electronic or optical rules constitute the *data link layer.* In other words, the link is a homogeneous physical span of the network, and the link layer comprises the rules for transmitting information over it.

For example, to get the signal from your household to the telephone central office in town, it must be modulated onto a three-kilohertz analog carrier. The rules for linking to the carrier and modulating it—all the hissing and beeping you hear when you connect your modem, for example—are part of the *data link layer.* Although this layer is not as fragmented and various as the physical layer, it will only get you across one path of a specific kind. The link changes again at the central office or Internet hub, where it may be converted into digital form, usually SONET frames of 64 kilobits per second or Ethernet packets of up to 1,550 bytes, or asynchronous transfer mode (ATM) cells of 53 bytes. These are all to be switched onto the path to the next link, which could be a phased array microwave antenna connecting to a satellite, or a messenger on a bicycle. All these connections, conversions, switching functions, and signaling rules comprise the *data link layer.* The key is that there are lots of these links, each dependent on a particular technology, and no one of them reaches from end to end, from origin to destination of the call.

If there were just one link—as in an original telegraph system—the physical and data link layers would be essentially all there was for manufacturers to supply with equipment. One company, such as Western Union (or AT&T), could do it all and InterOp would be a boring place. But to get an end-to-end connection in a modern global system, you need some higher level of abstraction that can ride on all the different links. You have to ascend to the *network layer.*

In the telephone system, the network comprises, among other things, the phone numbers, the voice coding, and the rules for maintaining circuits from end to end across the network. On the Internet the *network layer* begins in the TCP/IP protocol stack in your computer. TCP/IP stands for Transmission Control Protocol-Internet Protocol. The network layer is mostly IP, with packets led by a header of 32 bytes and holding a payload of up to 65 kilobytes. IP defines the hierarchy of network addresses and the routes to all the other routers. All these specifications of an end-to-end communication are contained in an IP packet header, which is a kind of electronic envelope. It contains an address, priority (air mail, special delivery?), data lifetime in total number of hops permitted (dead letter disposal), and other information needed to send messages across the network (or throw them away if they are not received).

The pivotal devices in the Internet's *network layer* are routers. Each router along the path contains a routing table in about 100 megabytes of memory. This holds all addresses on the router's own immediate network or subnet, as well as the paths to all other routers on the Internet. In a real sense, routers are the network, since they alone coherently deal with the totality of the end-to-end paths that constitute layer three. While switches in the link layer (layer two) only connect to the next link, routers weave together all the links. If it is not reachable through a router's lookup table, it is not part of the network. If a device contains all the end-to-end routes, it is not a switch, it is a router (though this nomenclature is often confused in advertisements for devices that combine different layer functions).

The *network layer* defines the entire network. However, you will probably only want to send particular messages over it. That will require you to make a specific connection through the *transport layer* which sets up the rules for a particular channel between the origin and the destination. TCP (Transmission control protocol) functions on the transport layer to control and balance the streams of packets across an IP network. Flow control is the name of this function. For example, TCP requires an acknowledgment of each packet. If packets are dropped, it slows its rate of transmission in order to prevent further congestion.

Transport defines a particular connection, but it does not control the actual session—how it begins and ends and how the dialog is managed. That is the job of the *session layer*. Moreover, some communications may comprise several connections. For example, a teleconference may include a number of transport linkages. A video teleconference may combine an image with a voice message and documents. The *session layer* handles these multiplicities.

The most basic network function, switching, happens back at layer two (datalink). It just moves the bits to the next hop. In a homogeneous network such as a campus LAN, switching might suffice. But switching cannot handle an Internet, with its chaos of subnets and suppliers. Routing—the more elaborate and software-intensive way of transferring Internet Protocol (IP) packets, using a lookup table of best routes rather than switching's simple hops—happens at layer three (network). The most embattled issue in networking today is whether to try to route at layer two by jazzing up a switch (with end-to-end virtual circuits, but never mind) or whether to switch at layer three (with a hardware "finite state machine," but again, never mind). In any case, layer three seems to be subsuming layer two. Or is it the reverse?

After memorizing the five bottom layers, you still need *p* and *a*, as in *chutzpa,* to really make it as a networking expert. The two final ones are for presentation and application layers, which actually originate or receive the message and constitute the actual functions being performed—E-mail, voice, file transfer, fax, transactions, images, trace routes, streaming media, plain or encrypted text. To most of the equipment suppliers, that's the frosting on the cake. To Internet users, it is the cake.

Having learned all these layers, you must accept that in practice they often give way to such further refinements as an encryption layer, a media access layer (MAC, located between two and three) along with several other "middleware" schemes designed to link IP and other protocols gracefully to physical media. If you lack the chutzpa for fiddlenets, of course, you can just go directly to the corporeal and spiritual layers at an InterOp hotel bar.

Don't forget the OSI layers, though, for they make the Internet possible. Without them, every supplier of equipment would have to comprehend all the functions and codes and polyglot interactions of the entire system. The layers function like parentheses in algebra. They isolate particular functions from the rest of the calculation. They allow specialization and modularity, increasing the speed of learning and enhancing the state of the art. Without the layers, the Internet would advance no faster than the old public switched telephone network (PSTN). Bill Joy of Sun Microsystems has ordained a law: "Most of the smartest people are never in your own company." By allowing smart people all across the net to innovate equipment for it, the layer model enables the astonishingly fast ascent of Internet technology.

Because of the layers, InterOp is more fraught with real debate and emotion than a political convention. Even though everything above the physical layer is nonphysical—software, code, rules, or logic—the logical functions are executed in physical machines. Changes in the logic can add, change, or eliminate machines, including multimillion-dollar state-of-the-art boxes. At the center of most arguments at InterOp and across the industry is the issue of whether the seven layers are too many or too few. Adding a layer—of middleware, encryption, or access, for example—often rings cash registers, or brings initial public offerings and new lucrative niches for the companies that do it. Removing a layer normally means amputating the hundreds or even thousands of companies that supply gear connecting at that level, hence the emotions.

If you want to see real emotions, talk about the telecosm revolution to InterOp companies. Because now that you've learned the seven-layer acronyms, you might just be able to forget about them. The fibersphere aims to eliminate virtually everything but the physical layer from the center of the network. Thus it remains highly controversial at InterOp, where most of the expertise revolves around complex protocols, often devoted to guaranteed "quality of service."

In the end, the fibersphere will render InterOp itself less and less significant. Indeed, in 2000 InterOp showed signs of a sharp decline, as many of the leading optical companies stayed away. They chose to focus instead on ascendant optical conferences, such as OFC (optical fiber conference), which concentrate on what they call the optical layer, leaving everything else as peripheral.

The seven-layer model epitomizes the *ancien regime*'s approach to communication networking. Make the network more and more intelligent, so it can connect more and more users, with proliferating kinds of equipment. Unleash a Brahmin class of engineers who preside over the center of the system, telling all of the Untouchable users what they can and cannot do.

To be sure, the intelligence at the center, among the seven layers, or even more layers as things get more complicated, is fearsome. But ultimately the Brahmins are no match for the revolution of the Untouchables. *Their* intelligence, distributed around the end points of the network, will take control. With infinitely abundant bandwidth, you can replace the seven-layer smart network with a much faster, dumber, unlayered one. Let all messages careen around on their own. Let the end-user machines take responsibility for them. Amid the oceans of abundant bandwidth, anyone who wants to drink just needs to invent the right kind of cup.

Chapter 7

The Law of the Telecosm

Among Netheads and Bellheads, the battle continues. The bandwidth tide is relentless, and it assures the triumph of dumb centers and smart edges. But the Kings of Bell are still fighting for centripetal intelligence, smarts at the center, high IQs assuring that every signal gets intelligently through, and every bit always knows its place. The seven layers of the netplex break into two polar paradigms. Most familiar is the public switched telephone network (PSTN in the obligatory acronym). From the tiniest transistor flipflop on a modem chip through labyrinthine layers of rising complexity on up to a 4ESS supercomputer switch linking 107,520 telephone trunk lines and itself consisting of millions of interconnected transistors, the PSTN is a vast, deterministic web of wires and switches. Once you are connected in the PSTN, your message is guaranteed to get through.

In the public switched telephone network, bandwidth constantly expands as you rise in the hierarchy. At the bottom are the twisted-pair copper wires of your telephone that function at three kilohertz (thousands of cycles per second). At the top are fiber-optic trunk lines that function at rates some multiple of the ultimate gigahertz speeds of the electronic transistors that feed the glass wires.

Peter Huber has described the five tiers of the telephone switching system as a structure with "the solidity, permanence, and inflexibility of the Great Pyramid of Cheops, which on paper it resemble[s]." It commands an enormous physical layer of forty-five million tons of twisted-pair copper wire at the bottom, rising up through a fiber backbone and the regal ranks of mainframe computers reduced to the tedious prestidigitation of switches. Utterly deterministic—there is a known path for everything—the PSTN's supreme reliability comes at the cost of inflexibility and huge inefficiency.

Most of its potential capacity—its bandwidth, in other words—is wasted most of the time. For example, until recently, the average U.S. home phone line was used only twenty minutes a day; 99 percent of the time it lay fallow. Even when it is being used, the PSTN wastes most of its bandwidth. The network can pass no information at all until it has created a complete end-to-end circuit connection dedicated to a particular call. During the silences in a conversation, the particular circuit is unavailable to any other user. The telephone is still largely a hit-or-miss medium. Fifty-five percent of all calls fail to reach their intended recipient. Progress has been glacial. Until the all-out assault of 1999 by competitive carriers, average long-distance prices scarcely dropped at all in nominal terms in twenty years. Over the last five years, they have declined just 50 percent adjusted for inflation, despite the fact that the physical system consists mostly of electronics governed by Moore's law. Some 40 percent of the cost comes from the billing systems themselves, administering cumbersome and exploitative practices that charge by the minute and gouge loyal customers who fail to shop for the best deal.

The key problem of the PSTN is that it will bring us neither the telecosm of bandwidth abundance nor the fibersphere that makes it possible. The network was designed so that the people who run it dictate how it can be used. This is not because telco executives are dictatorial by nature, but because the system was formed to serve dumb and narrowband voice terminals—telephones—with what they call five nines of reliability (essentially 100 percent). For this purpose, the PSTN is supremely intelligent and robust. For any other purpose it is paralytic.

David Isenberg has described the problem from inside. Working in vocal technology at Bell Labs, he had participated in the team that developed True Voice, a vaulting improvement in AT&T's voice quality in response to Sprint's arresting boast that you could hear the "sound of a pin dropping" over its fiber network. True Voice was a technical success, but the project provoked Isenberg to question the most fundamental assumptions of the AT&T network.

As he later explained, "True Voice was a valiant attempt to improve circuit-switched voice quality as much as possible in the context of current network architecture. If we had not been constrained by network architecture, the easiest way would have been to increase the sampling rate or change the coding algorithm." These changes could have been done simply at the terminals of a dumb network. For example, a psycho-acoustics expert recommended that AT&T boost the base frequencies it transmitted in order to make voices more lifelike. "But to actually do this, we would have had to change every piece of the telephone network except the wires," he said. Modems, fax machines, echo cancellers, voice mail systems all depended on "intelligent" assumptions about the acoustics of the signal that would be extremely expensive to alter.

As time passed, Isenberg became increasingly rebellious. He would dress up as a clown and organize seminars on such heretical cheap bandwidth topics as "What Would Happen If Minutes [Bandwidth] Were Free." Finally in June 1997, he penned a concussive essay, "The Rise of the Stupid Network," that sent seismic waves through the telco establishment and illustrated the power of an idea unleashed on the Internet, where his bosses had innocently permitted him to "hide" it.

One way or another, the world flocked to Isenberg within hours of the posting of his article. Harry Newton, the brash and brilliant computer telephony impresario, asked to publish it in his magazine, *Computer Telephony,* and Isenberg said no. Newton speaks only in a kinky Australian dialect and perhaps there was some part of Isenberg's "no" that he didn't understand. So Isenberg's article appeared in the August 1997 issue. The essay made him famous (or notorious) in the industry and impelled us to meet. He had no idea that I had been writing about dumb networks for years. But the data had led both of us to the same observation: that a bandwidth explosion was overthrowing the telephone company paradigm of scarce wires. A new paradigm of abundance dictates a new network architecture. But the system has been deeply resistant to change. As Isenberg put it, "Telcos invented the stored program control switch in the seventies," affording some flexibility, "and then fell asleep at this very switch."

The transistor and the laser, both invented at Bell Labs, will bring the system down. They both enable and require a dumb, all-optical network. But until the last few years, under CEO Michael Armstrong, few of the innovations of Bell Labs had been incorporated in to the telephone system. With its deterministic end-to-end circuits, its quality of service guarantees, its labyrinthine intelligence, the Bell establishment remains mostly in place today: the PSTN pyramid.

That is one network paradigm. The other paradigm is Robert Metcalfe's. It germinated in his mind in 1970 as he read a paper by Norman Abramson of the University of Hawaii. Abramson told of another paradigm. He called it Aloha. With Aloha, there were no guarantees.

The name comes from Alohanet, a packet radio system used for data communications among the Hawaiian Islands. Packets are collections of bits led by a header bearing an address; they proceed through a communications system rather like envelopes through a postal system. The key feature of Aloha was that anyone could send packets to anyone else at any time. You just began transmitting. If you didn't get an acknowledgment back, you knew that the message had failed to get through. Presumably your packets had collided with others. In Metcalfe's words, "They were lost in the ether." At that point, you would simply wait a random period (to avoid a repeat collision as both parties returned to the channel at once). Then you would retransmit your message.

To Metcalfe, Aloha seemed a beautifully simple network. But Abrahamson showed that because of collisions and other problems, it could exploit only 17 percent of its potential capacity. A student of computer science at Harvard searching for thesis ideas, Metcalfe believed that by using a form of advanced mathematics called queuing theory he could drastically improve the performance of Aloha without damaging its essential elegance and simplicity. What Metcalfe, then a twenty-six-year-old graduate student, eventually discovered would bring such networks up toward 90 percent of capacity and make Aloha a serious threat to the entire structure of the PSTN pyramid. By not setting up an end-to-end circuit for every call, the Aloha model can operate at hugely higher efficiencies than the PSTN, and technological advances, chiefly in optics, may well end up by rendering it more reliable as well.

Metcalfe's discovery is known as Ethernet. Twenty-five years later, Ethernet is the world's dominant local area computer network (LAN) and at fifty-one, Metcalfe is known and celebrated as its inventor. He is also founder of 3Com, the leading producer of Ethernet adapter cards and a major communications products company. In this era of networking, he is the author of what I will call Metcalfe's law of the telecosm, showing the magic of interconnections: *The value of a network rises in proportion to the power of all the machines attached to it.* Think of phones without networks or cars without roads and you can imagine the benefits of linking up computers, and sense the exponential power of the telecosm.

The power of the telecosm reproduces on a larger scale—interconnecting computers—the exponential yield of the microcosm interconnecting transistors on individual slivers of silicon. In the semiconductor industry, the measure of transistor efficiency is the power-delay product, a measure combining the switching speed or delay with the power usage or heat dissipation. As increasing numbers of transistors are packed ever closer together on individual chips, their power delay product improves exponentially; they function faster, cooler, cheaper, and better. Metcalfe's law suggests that a similar spiral of gains is available in the telecosm of computer communications.

As the creator of the system that links more than 80 percent of the world's one hundred twenty million networked computers, Metcalfe is still extending his paradigm and his law. Indeed, the law would suggest that in addition to his some $100 million of personal net worth from 3Com, he has fostered hundreds of billions of dollars in global wealth. Led by Cisco, the manufacturer of routers and related gear, with a market capitalization of some $300 billion in 2000, the top fifteen publicly traded computer networking companies have a total market value of some $300 billion. Add to that sum the value added to the world's two hundred million computers as they are increasingly linked to the Internet and you may sense the power of the Metcalfe paradigm.

Today, more than twenty-five years after he conceived it at Xerox's Palo Alto Research Center in 1973, Ethernet is still gathering momentum, gaining market share, and generating innovations. As Ethernet spreads and faces the challenge of remote work teams using digital images, simulations, maps, CAD schematics, visualizations, hi-fidelity sounds, and other exotic forms of data, the system is constantly adapting. It is emerging in full duplex, multimedia, fast, fiber-optic, shielded, unshielded, twisted, thin, thick, hubbed, collapsed, vertebrate, invertebrate, baseband, broadband, pair, quartet, coaxial, and wireless versions. It now can run at 2.9, 10, 20, 100 and 1,000 megabits a second. 10,000 megabit products are coming to market. In the last fifteen years, it has moved from 2.9 megabits per second to 10 gigabits per second and from a few hundred to more than a million users. At its present pace of progress, Ethernet will someday run isochronous (real-time) gigabits per second on linguini.

The ultimate power of the Ethernet model is manifested by the Internet, which largely comprises a group of local Ethernets connected together. Ethernet is a layer-two protocol, originally dealing with links on local area networks rather than with heterogeneous worldwide webs. But in a simpler form—call it an ethernet with a small "e"—it embodies the same concepts as the Internet. Both focus intelligence and control on the edge rather than in the center. Both overcame more intricate, more feature-rich, more centralized, and higher "quality of service" rivals. Complex in its terminals and simple in the core, connectionless and packet based, the Internet is largely an Ethernet writ large at the network layer.

Inspired by a radio network, Ethernet is well adapted both for fibersphere and atmosphere, both for the boundless bandwidth of glass and for the liberating world of wireless. The increasing movement of data communications into the air—the real ether—will give new life to Metcalfe's media-independent system. Cellular systems already operate with protocols similar to Ethernet. As microcells fill up with digital wireless traffic, linked to the Internet, all networks will increasingly resemble the Internet, be the Internet. In the ether, links will resemble ethernets far more than centralized telecoms.

Whether in glass or air, the basic protection of ethernets and internets is not smarts but statistics. Both are probabilistic systems. This fact has caused endless confusion. Because a probabilistic system cannot guarantee delivery of data on a specific schedule, or at all, many experts have concluded that Ethernet is unsuited for critical functions, or for isochronous data, such as voice and video, that must arrive in real time. When and whether anything arrives is a stochastic matter.

Nonetheless, if there is enough bandwidth for the application, ethernets work just as reliably and well as their deterministic rivals, even for advanced video traffic. If bandwidth is infinite, delays are limited to the speed of light, massless photons collide without clashing, and reliability is perfect.

From the beginning, bandwidth was the ultimate guarantor and promise of the Ethernet. Its early capacity of ten megabits per second far outpaced the powers of all but mainframe computers to fill. Now "fast Ethernet" at 100 megabits per second vastly outreach the needed bandwidth of desktop machines, and 10 gigabit Ethernets overprovision all but the most video-intensive networks. The age of bandwidth abundance is a supreme vindication of the Ethernet vision. So why is its still boyish-looking inventor—think of Ted Kennedy some ten years ago—giving up on his baby just as it swings triumphantly through its roaring twenties? Why is he ready to abandon his basic paradigm in favor of a return to the PSTN vision of massive intelligent switching systems at the very moment that the Internet is sweeping all before it as the apotheosis of the Ethernet vision? Why is Metcalfe now talking of Ethernet as a "legacy LAN"? Why did he predict in 1996, in a famous column that he promised to eat if he was wrong, the crash of the Internet by the end of the year?

He began expressing qualms about the triumphal trend in networks in the early summer of 1993, discoursing from a deck chair on his yacht (which motored him down from his summer home in Maine to a dock on the Charles for his twenty-fifth MIT reunion). Metcalfe has the air of an elder statesman, humbly grateful for the benisons of Ethernet. But he has seen the future in a poll of experts prophesying the universal triumph of a powerful new switching system called asynchronous transfer mode (ATM). "I have found," he solemnly intoned, "an amazing consensus among both telephone industry and computer networking experts that ATM is the future of LANs." Aloha, ATM.

Although this ATM has long been expected to gush jackpots of cash for gaggles of network companies and investors, it is unrelated to automatic teller machines. Think of ATM rather as an automated postal center that takes messages (of any size or addressing scheme), chops them up, puts them into standardized little envelopes, and sets up routes to their destinations in billionths of seconds. The magic of ATM comes from restricting its services to those uniform envelopes (called cells) of 53 bytes apiece (including a 5-byte address) and creating what is called a virtual circuit connection through the network for each flow of related envelopes (such as a group comprising an image or a voice signal). These features make it unnecessary for intermediate switches in the network to check the entire address; the flow of cells flash through the system on a precomputed course.

Perhaps most attractive of all, ATM can handle multimedia data, such as digital movies or teleconferences, with voice, text, and video that must arrive together at the same time in perfect sync. As the world moves toward multimedia, the industry flocked toward ATM, the innovation that can make it possible.

By contrast, Ethernet seems old and slow: the vacuum tube of computer communications. Think of the first versions, crudely, as a system where all the messages are cast into an ocean and picked up by terminals on the beach, which scan the tides for letters addressed to them. Obviously this system would only work if the beach terminals could suck up and filter tremendous quantities of sea water. The magic of Ethernet comes from the ever growing power of computer terminals. The microcosm supplies sufficiently powerful filtering chips—chiefly digital signal processors improving their powers some tenfold every two years—to sort mails and messages in the vasty deep. This is quite a trick. To the experts, it seems unlikely to prevail for long against the fabulously swift switching of ATM.

So why is Metcalfe suddenly deferring to experts? As he knows all too well, if the experts had had their way, there would never have been an Ethernet. The experts at Harvard initially even rejected his thesis. Sitting with Metcalf on a bench at the Washington National Airport, Leonard Kleinrock, the king of queueing theory, spurned a key Metcalfe paper on Ethernet queueing, as Metcalfe recalls, because his word-processing program didn't have Greek letters for the math. Other experts banished Metcalfe from the Ethernet standards meetings because of wrong credentials. In 1986, when IBM introduced its new Token Ring network system—far smarter, 60 percent faster, and containing eight classes of quality-of-service guarantees—most of the experts predicted the death of Metcalfe's baby. But Ethernet overcame all odds and Token Ring is now dying a long, slow death. Metcalfe made his entire career defying the experts. Now he is joining them. Why?

Part of the problem may be that Metcalfe, now publisher and chief pundit at *InfoWorld,* became something of a credentialist himself after finally extorting a doctorate in computer science from Harvard. Of course, in the matter of relevant expertise, a computer doctorate from Harvard is rather like a doctorate from Caltech in astrology, if there were such a thing . . . or even a business management degree from MIT (during his nine years of matriculation, Metcalfe also picked up one of them). Even though Metcalfe offers fervent assurances that he never stepped foot on the Harvard campus after gaining a master's in applied math there—researching his computer thesis mostly on the MIT campus and at Bolt, Beranek & Newman—the man is hardly one of those dropout entrepreneurs that he calls his spiritual kin.

In his book of 2000, written in some of the crispest prose ever penned by an engineer (believe it or not, Metcalfe is also an MIT double E), he gives a vivid account of the decision by experts at General Electric to shun the personal computer and networking businesses. Meeting in the fall of 1980 at the very dawn of the personal computer era, they deferred to skeptical focus groups despite strong recommendations from Metcalfe and his newly formed 3Com.

Composed of assemblages of "ordinary GE customers" across the country, the focus groups concluded that there could be no large market for personal computers and networks. Computers would always be a specialized business market and what went for computers was still more true of computer networks.

The high point of the videotaped GE presentation was testimony from a mathematics teacher in Waltham, Massachusetts—the focus group's sole advocate of PCs—who climaxed a visionary celebration of computer networks by declaring: "In short, I see home computers as the next major new home appliance . . . I see companies like GE selling computers as they now sell refrigerators . . . by the millions . . . I see computers on the verge of changing our entire concept of home and family."

Here the man paused, looking straight into the camera. As Metcalfe put it, "He clearly had one sentence to go."

The man uttered it with fervor: "And, I don't mean just here on earth."

GE decided against computers and networks—and later against chips and software as well—and became essentially a portfolio of the kind of companies beloved by focus groups: firms such as the NBC network, the railroad turbine business, "white" goods, medical technologies, and military hardware which the relevant customers already understand. GE thus shunned what is today the world's largest industry.

At the same time, Bob Metcalfe resolved to launch his company 3Com into the business of developing computer networks for businesses, "just here on earth." Early in 2000, this limited goal brought 3Com to a market value of $11 billion.

Nonetheless, let us not forget our Waltham transcendentalist. Nearly everything the math teacher foretold has indeed come true. If Metcalfe had kept his head in the ether with the man from Waltham he might have better understood why Ethernet not only is the dominant network of the past but also will be dominant in the future too. ATM—like Token Ring and solar energy—will obviously find niches as the rave of the experts. But Ethernet is quietly preparing for a new era of hegemony in the marketplace for computer connections.

Ethernet is still selling 24 million installations a year, compared to Token Ring's annual total of under a million. But this does not persuade Ethernet pioneers Metcalfe and Kleinrock. Because ATM can handle all kinds of data fast, Metcalfe sees it as the "grand unifier," bringing together WANs and LANs and effecting a convergence of television, telephony, and computing in turbulent multimedia bit streams bursting into our lives. "And of all the variations of multimedia," he writes, "the one that will drive ATM is personal computer videoconferencing—interactive, two-way, real-time, integrated digital voice, video, and data." Although Ethernet will persist as a "legacy LAN," he says, it cannot compete with ATM in these crucial new roles.

Bringing mathematics to bear on the argument, Kleinrock declares that the collision-detecting functions of Ethernet bog down with large bandwidths, short packets, and long distances. Thus the system must fail with the onset of the fibersphere. The oceans of ethernet will simply grow too large to allow efficient detection of collisions in its depths. When the distances get too long, collisions can occur far from the transmitting computer and take longer to be detected. The shorter the packets the worse these problems become.

As Kleinrock computes these factors, the efficiency of Ethernet is roughly a function (A), computed as five times the length of the line in kilometers times the capacity of the system in megabits per second divided by the packet size in bits. When A exceeds a certain level (Kleinrock sets it at .05), Ethernet's efficiency plummets.

With ATM packet sizes needed for voice traffic—or even at the minimum ethernet packet size of 72 bytes—any Ethernet with a capacity much higher than 10 megabits per second exceeds this tipping point. Therefore, high-speed Ethernets must either use packets too long for voice or shrink in extent to far less than 3 kilometers.

But to the experts, short-link Ethernets fail to address the multigigabit world of fiber optics. At some point, Kleinrock and Metcalfe agree, ad hoc fixes will begin to fail and ATM will begin to prevail. Using Kleinrock's formula, that point has supposedly already arrived with 100 megabits-per-second Ethernet lines. In a world of multimedia, the triumph of ATM, so it would seem, is just a matter of time.

Time, however, is precisely what is absent in all these projections. Ethernet is a system indifferent to the media it uses. When fiber came along, it merely allowed the increase of Ethernet to gigabit levels while increasing the distance it could carry to hundreds of kilometers. Ethernet is a system based on the intelligence of terminals; ATM is a system based on intelligence in switches and networks. The near annual doubling of chip densities and the spiraling increase of computer power surging on the fringes of all networks has made Ethernet increasingly powerful as the power of computers soars.

Amazingly, most technology prophets fail to come to terms with the power of exponents. You double anything annually for long—whether deforestation in ecological nightmares or transistors on silicon in the awesome routine of microchip progress—and you soon can ignite a sudden moment of metamorphosis: a denuded world or a silicon brain. Looming intelligence on the edge of the network will relieve all the current problems attributed to ethernets and will render the neatly calculated optimizations of ATM irrelevant.

With terabits per second running over continents and under oceans through all optical lines, the numerator in the Kleinrock formula is rising by a factor of millions, rendering irrelevant the entire efficiency calculations. If you have enough bandwidth, you don't have to worry about collisions. This bandwidth abundance makes it nothing short of ridiculous to expect a system optimized for 1997 chip densities and fiber capacities to remain optimal in 2013, when Metcalfe foresees the final triumph of ATM, or even in 2003.

The lesson of IBM's Token Ring returns to overthrow ATM. Although Token Ring was built with eight guaranteed levels of quality of service, virtually all packets ever sent through token ring networks ran at the highest QoS (quality of service). The eight levels merely added complexity to the system, but all the users chose first class. Today on the Internet, the consensus claims that QoS will be indispensable for voice and video. But with true bandwidth abundance, QoS complexities are irrelevant—an ATM tax imposed on the vast bandwidth of fiber with its 10 to the minus 15 error rates, far better than the reliability of telephone circuits.

In 1973, Metcalfe could not have anticipated the stunning miracle of fiber optics, with a thousand wavelengths on a single fiber thread, 864 fibers in a single cable, and a feasible capacity of some 8.6 petabits a second. He could not have foreseen a world of bandwidth abundance in which communications power could be wantonly wasted. He could not have grasped the increasing possibility—everywhere except on the farthest edges of the net—of a return to circuit switching based on wavelengths. But none of that mattered. For he understood vividly that the crux of his system was to refrain from compromising the future by dedicating his Ethernet model to any single medium.

This Metcalfe vision is imperative today. The combination of intelligence at the terminals and statistics in the network is more robust than the mechanistic reliability of Token Rings, ATM switches, or any other quality of service requirements. As Metcalf points out, in explaining the triumph of his vision over Token Ring, Ethernet is a simple system that is stabilized by its own failures. The CSMA/CD algorithm uses collision detection in a negative feedback loop that delays retransmission in exponential proportion to the number of collisions, which is a reliable index of the level of traffic. Similarly, TCP, the Internet flow control protocol, uses dropped packets as a signal of congestion and adjusts in real time. Thriving on worst-case assumptions of frequent failure, both Ethernet and the Internet have outpaced all rivals that guarantee perfect performance and depend on it.

Now in ATM, Ethernet is faced with a new paragon of determinism offering high speeds and rigorous guarantees, a new version of the public switched telephone network paradigm, a new pyramid of switching power. But Metcalfe's law and legacy will win again, in spite of his own qualms.

As Metcalf explains: "Ethernet works in practice but not in theory." The same could be said of all the devices of the microcosm and telecosm. Both the supreme sciences that sustain computer and communications technology—quantum theory and information theory—are based on probabilistic rather than deterministic models. They offer the underpinnings for an age of individual freedom and entrepreneurial creativity.

Man's constant search for deterministic assurance defies the ascendant science of the era, which finds nature itself as probabilistic. To Einstein's disappointment, God apparently does throw dice. But chance is the paradoxical root of both fate and freedom. Both nations and networks can win by shunning determinism and finding stability in a constant shuffle of collisions and contentions in ever expanding arenas of liberty.

Because of an acceptance of setbacks, capitalist markets are more robust than socialist systems that plan for perfection. In the same way, successful people and companies have more failures than failures do. The successes use their faults and collisions as a source of new knowledge. Companies that try to banish chance by relying on market research and focus groups do less well than companies that freely make mistakes and learn from them. Because of an ability to absorb shocks, stochastic systems in general are more stable than determinist ones. Listening to the technology, we find that ethernets and Internets resonate to the deepest hymns and harmonies of our age.

Chapter 8

The Wireless New World

The miracle of the fibersphere is that it improves old copper bandwidth and error rates by a factor of billions—in Paul Green's memorable estimate by ten orders of magnitude. This ten-billionfold multiple is a quantity with qualitative clout—and the crucial factor to grasp in fathoming the new economy.

Fiber, though, is an immobile miracle. It is still linked to the computer age: the old frieze of desktop computers and tethers plugged into the wall. Entering the fibersphere you pay a toll. To use the broadband magic of the fiber-optic system, you have to find the right connection among scores of wavelengths in a tiny fiber core, one tenth the width of a human hair, and then link to it. That further miracle of photonics is a matter of socket science, requiring exquisite pains and precision.

You cannot do it on the road. You cannot do it while jogging jauntily along the Charles River or gasping for breath on a Grand Teton hike, or cruising the Santa Monica Freeway or turning onto the George Washington Bridge. You probably cannot do it while waiting to board your airplane at Ronald Reagan National Airport, or schussing with the digerati at Davos or reclining in a gondola on the Grand Canal. Fiber optics might enable you to visit all these sites through some kind of three-dimensional telepresence. Fiber baron Jim Crowe of Level 3 promises such experiences through the ultimate magic of what he calls "silicon economics," and you will be appropriately grateful to Crowe and his colleagues when they bring it about in silica. But being a human being, you will still prefer mobility much of the time, even if it means that now and again you end up like John Malkovich—muddy in a ditch beside the New Jersey Turnpike.

Without mobile access, the bounties of fiber optics will remain indoor pleasures for the plugged in and wired, couched and recumbent. Today, however, wireless is at last advancing in some ways as fast as fiber. The key to this success is the belated recognition that wireless technology is not fundamentally different from fiber technology. It is part of the same paradigm, with the same calculus of power budgets, bandwidth, wavelengths, and the speed of light. Think of electromagnetic radio waves as photons insulated by the atmosphere rather than by glass. As branches of spectronics, deriving from Maxwell's rainbow and information theory, these communication tools are technological kin and complements. Neither will finally prevail without the other. The fibersphere needs the atmosphere as your lungs need the air.

The most common personal computer is already a digital cellular phone, a teleputer. There are 300 million cellphones in use, increasingly digital ones, and just 200 million personal computers. Soon your cellphone will be more powerful than your conventional PC. It will be as portable as your watch and as personal as your wallet. It will recognize speech and convert it to text. It will plug into a slot in your car and navigate streets. It will collect your news and your mail and if you wish, it will read them to you. It will browse the World Wide Web and download information as needed. It will tell you the weather at your destination. It will conduct transactions and load credit into an encrypted credit meter on its microprocessor, or onto a chip on a smart card, which can be used like cash. It can pay your taxes, or help you legally avoid them. It will take pictures and project them onto a wall or screen, or dispatch them to any other teleputer.

It just may not do Windows. But it will do doors—unlock your car door, open your front door, unroll your garage door, or even break on through to the other side of the doors of perception with Jim Morrison, if you favor those swinging Doors of the 1960s. It will hold books and display them in high resolution or have them printed out at a convenient printer. It will have an Internet address and a Java run-time engine that allows it to execute any applet or program written in that increasingly universal software language. It will link to any compatible display, monitor, keyboard, storage device, or other peripheral through infrared pulses or radio frequencies. Plug a notebook computer into the cellphone or link them wirelessly and you can connect to the Web at speeds up to 2.4 Mb/s, well beyond the pace of your office T-1 line (1.544 Mb/s) and roughly forty-three times as fast as your 56 Kb/s dialup modem.

Faster and better than the wireline mean, these new wireless systems are largely fruits of a new concept in communications that has been developed and launched over the last decade by a company in San Diego named Qualcomm that few people had heard of until 1999, when its shares rose in value by 2,619 percent and it became the leading star of the globe's stock markets. Qualcomm could defy the world and prevail chiefly because it listened to the technology, while its rivals listened to their customers and their experts. Qualcomm moved with the power of the paradigm that its adversaries were resisting.

Like the computer establishment before it, current cellular providers are still ambivalent toward the new vectors of the technology. They still tend to dismiss Qualcomm as the proponent of a "religious war," as if old wireless concepts could have done just as well or better if Irwin Jacobs and Andrew Viterbi, the Qualcomm prophets, had been less greedy for technical elegance and personal credit.

Ready or not, though, the revolution will transform the landscape over the next decade. It will bring the new broadband paradigm to wireless networks, foster a new attitude toward spectrum, and consummate the fiber-optic promise.

Virtually all electromagnetic radiation can bear information, and the higher the frequencies, the more room they tend to provide for it. As a practical matter, however, communications engineers have aimed low, thronging the frequencies at the bottom of the spectrum, constituting far less than 1 percent of the total span.

The vast expansion of wireless now in prospect, however, will require an ascent to higher frequencies far up Maxwell's rainbow, even scaling the infrared radiations of fiber optics. The fibersphere and atmosphere will merge into a unified realm of communications.

• • •

So naturally does communication come to human beings that no one thought to provide a theoretical explanation for it until the late

1940s. Expounding the rules of the realm was a great man who walked the halls of Bell Labs in the late 1940s at the same time as future Nobel laureate William Shockley, and who also left the world transformed in his wake. In 1948, the same year that Shockley invented the transistor, Claude Shannon invented the information theory that underlies most modern communications and computer science.

Information theory calculates the ability to transmit information over a channel of a particular bandwidth in the presence of noise. With nearly no noise and nearly infinite bandwidth in the fibersphere today, fiber-optics engineers have finally entered a kind of Shannon nirvana. But Shannon illuminated every step of the way. Now in the noisy and narrowband arena of wireless, no one can move without consulting Shannon. He is the father of the communications revolution of this era. If his information theory breakthrough does not fit any of the existing Nobel categories—Shannon's relevance spans such far-flung fields as biology and economics—perhaps the committee can find a new category for an appropriate prize.

In defining how much information can be sent down a noisy channel, Shannon showed that engineers can choose between narrowband high-powered solutions and broadband low-powered solutions. On the assumption that usable bandwidth is scarce and expensive, most wireless engineers have striven to economize on it. Just as you can get your message through in a crowded room by talking louder, you can overcome a noisy channel with more powerful signals. Engineers therefore have pursued a strategy of long and strong: long wavelengths and powerful transmissions with the scarce radio frequencies at the bottom of the spectrum. They have used power as a substitute for bandwidth.

Economizing on spectrum, scientists focused on analog systems, such as conventional radio and television, and then extended them to wireless telephony. Nothing could be more spectrum efficient, so it would seem, than sending an exact simulacrum of the contents to be transmitted—an exact analog of sounds and images—and use every point on the wave to convey information. Because these signals offer an exact analog of an image or sound, however, any interference or "noise" appears as static or "snow." Therefore, analog engineers invoked the Federal Communications Commission to exclude all possible intruders from their frequency bands.

This long-and-strong approach, with exclusive use of spectrum bands, seemed hugely more efficient than digital systems requiring complex manipulation and conversion of long strings of on-off bits. Taking the dense and accurate information contained in an analog wave and reducing it to a mere on or off, one or zero, digital code, seems to the analog engineer to reek of waste.

Ironically, this policy of economizing on spectrum eventually led to using it all up. When everyone talks louder, no one can hear very well. Today, the favored regions at the bottom of the spectrum are so full of spectrum-hogging radios, pagers, phones, television, long-distance, point-to-point, aerospace, and other uses that heavy-breathing experts speak of running out of "air."

Anticipating this predicament, Shannon offered a new paradigm, redefining the relationship of power, noise, and information. He showed that a flow of signals conveys information only to the extent that it provides unexpected data—only to the extent that it adds to what you already know. Shannon termed this information content "entropy." In a digital message, another name for a stream of unexpected bits is random noise. Termed Gaussian, or white, noise, such a transmission resembles random "white" light, which cloaks the entire rainbow of colors in a bright blur. Shannon showed that the more a transmission resembles this form of noise, the more information it can hold, as long as it is modulated onto a regular carrier frequency. In the esoteric language of Shannon, you need a low entropy carrier to bear a high entropy message.

Shannon's alternative to long and strong is wide and weak: not fighting noise with electrical power but joining it with noiselike information, not talking louder but talking softer in more elaborate codes using more bandwidth. Shannon acknowledged that added power improves the efficiency of an analog signal—you can hear the unexpected sound more clearly if it is louder. But he showed that more power degrades digital efficiency. Digital signals come in on-off bits, binary yeses and noes. If you receive a bit at all, it doesn't matter how loud it is; a bit is a bit. But a loud signal can drown out the messages around it.

What matters for digital communications is bandwidth. In contrast to analog communication, which improves merely by the logarithm of the bandwidth, digital efficiency increases roughly by the square of the bandwidth. In other words, in Shannon's world of noisy channels—such as a wireless cell or a cable TV coax—a doubling of bandwidth does not merely double the amount of information that can be sent; it at least quadruples it. Here we have one of those exponentials, like Moore's law and Metcalfe's law, on which new technology paradigms can be built.

Telecosmic companies such as Qualcomm use Shannon's bandwidth as a replacement for power and a remedy for noise and interference, and are prevailing against all the powers and principalities of global telecom. The law of the telecosm dictates that the higher the frequency, the shorter the wavelength; the wider the bandwidth, the lower the power; the smaller the antenna, the slimmer the cell; and, ultimately, the cheaper and better the communication.

With virtually no notice in the media, the fight for the future has been raging furiously for a decade. Under pressure from EEC industrial politicians working with the guidance of engineers from the giant Swedish corporation Ericsson, the Europeans adopted a new digital cellular system originally called *groupe speciale mobile* (GSM) after the commission that conceived it. With more than 100 million users in 147 countries by 2000, GSM became a global standard and its acronym changed to global system mobile.

GSM is a conservative digital system that excels the number of users in an analog cellular channel by a factor of three. GSM uses an access method called time-division multiple access (TDMA). Reflecting the time-sharing methods of centralized computers with large numbers of users, TDMA stems from the time-division multiplexing employed by phone companies to put more than one phone call on each digital line. Thus, both the telephone and the computer establishments are comfortable with time division.

Under pressure from European firms eager to sell equipment in America, the U.S. Telephone Industry Association in 1991 adopted a TDMA standard (IS-54) similar to the European GSM. Rather than creating a wholly new system exploiting the advances in digital signal processors in Shannon's paradigm, the TIA favored a TDMA overlay on the existing analog infrastructure. McCaw (now AT&T), Bell South, and the then Southwestern Bell (now SBC) took the TDMA bait.

Thus, in the name of competitiveness and technological progress, and of keeping up with the Europeans and Japanese, the U.S. moved to embrace an obsolescent cellular system. Just as in the earlier case of analog HDTV, the U.S. government fell for foreign claims. And just as in the case of HDTV, the entrepreneurial creativity of the U.S. digital electronics industry launched an array of compelling alternatives just in time.

Infusing cellular telephony with the full powers of wide and weak are a group of engineers from MIT who worked at Linkabit, where they originally developed key patents in TDMA. Then they spun out to launch a new company, Qualcomm, to replace their invention. Founded by former professor Irwin Jacobs and telecom pioneer Andrew Viterbi, Qualcomm has become one of the greatest entrepreneurial stories in the history of enterprise. The leaders of Qualcomm received the ultimate accolade for an innovator: They were all told their breakthroughs were impossible.

Rather than compressing each call into between three and ten tiny TDMA time slots in a 30-kilohertz cellular channel, Jacobs and Viterbi developed a system called CDMA (code division multiple access) that differentiates calls by multiplying their digital signal with a code that resembles white noise. It spreads a signal across a comparatively huge 1.25-megahertz swath of the cellular spectrum. This allows many users to share the same spectrum space at one time. Each phone is programmed with a specific pseudonoise code, which is used to stretch a low-powered signal over a wide frequency band. The base station uses the same code in inverted form (with all the ones turned into zeros and all the zeroes into ones) to "despread" and reconstitute the original signal. All other codes remain spread out, indistinguishable from background noise.

Jacobs compares TDMA-GSM and CDMA-Qualcomm to different strategies of communication at a cocktail party. In the TDMA analogy, each person would restrict his talk to a specific time slot while everyone else remains silent. This system would work well as long as the party was managed by a dictator who controlled all conversations by complex rules and a rigid clock. In CDMA, on the other hand, everyone can talk at once but in different languages. Everyone listens for messages in their own languages and ignores all other sounds as background noise. Although this system allows each person to speak freely, it requires constant control of the volume of the speakers. A speaker who begins yelling can drown out surrounding messages and drastically reduce the total number of conversations that can be sustained.

For years, this problem of the stentorian guest crippled CDMA as a method of increasing the capacity of cellular systems. Spread spectrum had many military uses because it was very difficult to jam or overhear. In a cellular environment, however, where cars continually move in and out from behind trucks, buildings, and other obstacles, causing huge variations in power, CDMA systems would be regularly swamped by stentorian guests. Similarly, nearby cars would tend to dominate faraway vehicles. This was termed the near-far problem. When you compound this challenge with a static of multipath signals causing hundreds of gyrations in power for every foot traveled by the mobile unit—so-called Rayleigh interference pits and spikes—you can comprehend the general incredulity toward CDMA among cellular cognoscenti.

Indeed, in 1991, leading experts at Bell Labs, Stanford University, and Bellcore confidently told me the problem was a showstopper; it could not be overcome. Critic Bill Frezza summed up the difficulty of the system: "Coordinating the real-time transmit power of hundreds of roving users . . . reminds me of the guy on *Ed Sullivan* who used to spin dozens of plates on bamboo poles: Turn your back for a moment and you're screwed."

Radio experts, however, underestimate the impact of the microcosm. Power is a homogeneous force fully understood by engineers. Controlling it is in fact much simpler than moving among frequencies and shuffling time slots for many varieties of data in TDMA systems. Using digital signal processing, error correction, and other microcosmic tools, Qualcomm's engineers surmounted their power problem with a series of feedback loops. With automatic gain control in the handset and constant surveillance of signal-to-noise and error rates from the base station, wattage spikes and pits are regulated by electronic circuitry that adjusts the power at a rate of more than eight hundred times a second. This is a fancy way of saying that Qualcomm's phones and relay stations constantly equalize the volume of the voices in each cell in order to focus on the one Swahili speaker at the party of a thousand guests.

Not only is it simpler than shuffling time slots, this power-control mechanism has the further effect of dynamically changing the size of cells. In a congested cell, the power of all phones rises to overcome mutual interference. On the margin, these high-powered transmissions overflow into neighboring cells where they may be picked up by adjacent base station equipment. In a quiet cell, power is so low that the cell effectively shrinks, transmitting no interference at all to neighboring cells and improving their performance. This kind of dynamic adjustment of cell sizes is impossible in a TDMA system, where adjacent cells use completely different frequencies and fringe handsets may begin to chirp like Elmer Fudd.

Once the stentorian voice could be instantly abated, power control changed from a crippling weakness of CDMA into a commanding asset. Power usage is a major obstacle to the PCS future. All market tests show that either heavy or short-lived batteries greatly reduce the attractiveness of cellphones. Because the Qualcomm feedback system keeps transmit power always at the lowest feasible level, batteries in CDMA phones last longer than in TDMA phones. Transmit power comprises some 25 percent of the power usage in a handset, and voice activity is another 30 percent. Unlike TDMA, CDMA inherently captures for other uses every pause or silence in the conversation.

A further advantage of wide and weak comes in handling multipath signals, which bounce off obstacles and arrive at different times at the receiver. Multipath just adds to the accuracy of CDMA. The Qualcomm system combines the three strongest signals into one. Called a rake receiver and coinvented by Paul Green of fiber fame, this combining function works even on signals from different cells and thus facilitates hand-offs. In TDMA, signals arriving at the wrong time are pure interference in someone else's time slot. In CDMA, which does not chop up the signal into time slots, the multipath signals come in time to strengthen the message.

Finally, CDMA allows simple and soft hand-offs. Because all the phones are using the same spectrum space, moving from one cell to another is relatively easy. Qualcomm has reduced all the digital signal processing for CDMA into one application-specific chip. Supplanting the multiple radios of TDMA—each with a fixed frequency—are digital/signal-processing chips that find a particular message across a wide spectrum swath captured by one broadband radio.

CDMA is not merely a clever way of sending bits through the air; it is a new spectronics paradigm. The new paradigm sees the use of abundant bandwidth, at ever higher frequencies, to compensate for newly scarce power and switching.

Urgently needed for a decade has been a wireless telecom strategy and technology adapted to the pattern of abundance and scarcity of the new era. Ultimately it should offer service as cheap and clear as wireline, far more convenient, and readily usable for data. It should join bandwidth abundance, gained through moves up spectrum, with a recognition of power scarcity as the prime problem of wireless, and it should focus on Internet data as the dominant market.

Incorporating the TCP/IP Internet protocols in every handset, Qualcomm's CDMA is the first such system. Because it threatened the old order, it evoked more furious and bitter resistance than any new technology since the VCR. As a speaker around the world, I have faced the full fury of the resistance to this technology. As the author of books against feminism, I have met many hostile audiences in my time, but none raged as hotly as European telecom executives against CDMA.

The virtues of the system melted little ice among the proponents of GSM. As late as September 8, 1997, under the title "Blind Faith," *Telephony* magazine, the industry's leading American journal, ran what seemed a devastating expose of the scandal of American resistance to GSM. *The Wall Street Journal* in 1998 published a front-page story on Irwin Jacobs that treated his enthusiasm for CDMA as a kind of mental disease that impelled him to wild exaggerations and marketing hype. On January 24, 2000, *Fortune* magazine greeted the new millennium with a typically gullible account of the superiority of GSM. It decried the religious wars of American wireless, depicted GSM as "the most advanced digital standard to date," and declared that government guidance would give the Europeans third-generation (3G) phones by 2002 "with nifty new features," while "Americans probably won't have 3G networks until sometime between 2003 and 2005." Whenever I touted CDMA in Europe over the last six years, GSM proponents would tell me in all earnest that CDMA claims were contrary to communications theory principles and the rules of radio propagation. Even in the U.S., the rambunctious CDMA skeptic Bill Frezza accused the two key Qualcomm figures, Jacobs and Viterbi, both with EE doctorates, of wild mendacity for their supposed claims that CDMA could ultimately improve on analog by a factor of 20 to 40. (Don't look now, Bill, but it's happening.) The general implication was that these guys should be in jail for violating the laws of physics.

CDMA's vindication came first in 1997 in Seoul, Korea, where CDMA finally proved its superiority. Long the chief opponent of CDMA, Ericsson Corporation of Sweden gulped, launched a suit against Qualcomm, claimed to have invented CDMA in the first place, and ceded its superiority for data.

By 2000, Korea boasted more than 20 million CDMA phones and regarded CDMA equipment, produced by Samsung, LGS, and Hyundai, as a rare bright spot in a then struggling economy. Everywhere, CDMA was between three and six times more efficient than TDMA in the use of spectrum and hundreds of times more efficient in the use of transmitted power. Most important of all, CDMA continued its advance around the globe.

The key breakthrough—so little recognized that Qualcomm's stock rose just 3 percent during that year—came in January 1998 when the European Telecom Standards Institute (ETSI) endorsed CDMA as the next generation of GSM. Although the Europeans continued to put the best face on the capitulation, the triumph of CDMA was increasingly undeniable. Early in the year 2000, the International Telecommunications Union (ITU) in Geneva approved a set of five variations on a new standard for the third generation of wireless. All were based on Qualcomm's CDMA technology.

With the victory nearly total, I wanted to find out what the company planned for an encore, particularly with regard to Internet data, the obvious next frontier for wireless and a significant factor in the victory over TDMA. Owner of Eudora, Qualcomm was a leading E-mail company; I wanted to get my E-mail through my CDMA cellphone.

The future of wireless is Internet access. The purpose of third-generation phones is to combine voice and data. Not the Europeans but Qualcomm is well in the lead in this race. I got the chance to learn their secret on a trip to the company's headquarters north of San Diego in May 1998. A little early and a little groggy, I arrived at the offices and scanned the atrium of the polychrome hilltop palace that Qualcomm bought from the Fashion Institute when it overrated the willingness of San Diego architects and designers to move out to La Jolla. I needed a grande latte and it seemed a perfect place for a Starbucks. But instead I was shunted off to what I was told was a CDMA museum.

Curvacious but obsolescent telephones are not my thing, so I was pleased to discover that the "museum" was a small room filled with spectroscopes. It was manned by two old Qualcomm hands who just happened to be heroes of the history of spread spectrum radio. One was David Clapp, who worked under Klein Gilhousen, architect of the first CDMA prototypes in 1989. The other was Phil Karn, who had written key papers on spectral efficiency and channel access for spread spectrum radios and had engineered the inclusion of a TCP-IP Internet protocol stack in every Qualcomm phone way back in antediluvian 1991. This was a crucial moment in the history of cellular data.

Karn recalled, "I had previously written a stack for ham radios in late 1980. Then I came to Qualcomm and wanted to get it into the CDMA architecture. There were a lot of arguments, saying that TCP-IP was too complex for cellular and the Internet was only used by geeks. But I'd been involved in the net at Bellcore since 1985 and I knew it was growing exponentially. Exponential growth has a way of mounting until even phone companies can notice it."

Largely as a result of Karn's prescience, and the receptivity of his bosses, Jacobs and Gilhousen, Qualcomm phones would be able to link directly to the net from a laptop without a modem. Karn went to the Cellular Telephone Industry Association (CTIA) in 1993 in an effort to get this capability standardized. But he discovered to his surprise that telephone people *like* modems. They actively resist their replacement with direct digital connections. As an AT&T executive put it: "We don't like where this Internet stuff is leading." He explained it gave too much control to users, which from the point of view of a centrally controlled national network with essentially 100 percent uptime it did.

Qualcomm's new system, as deployed in Korea, fulfills Karn's early vision, combining the very different functions of data and voice. Using all the available spectrum all the time, CDMA is a shared medium like Ethernet that can handle sudden bursts of broadband data and narrowband voice in stride. The next Qualcomm pure-data system, called High Data Rate, transmits at up to 2.4 megabits per second. Superior to nearly all wireline modems, including most cable and DSL devices, HDR gives Qualcomm a chance to get its technology in every PC and network appliance. Future CDMA systems will combine data and voice at rates of hundreds of kilobits per second, fast enough for full motion video. Far from lagging the Europeans in 3G systems, the U.S. is well in the lead.

For many such reasons, the Europeans decided to go for CDMA-spread spectrum. Although conference wiseacres around the globe tout marketing or politics or monopoly lock-in as the key forces in business success, you had better bet on technology with your pension funds. In this case, from the EEC and the U.S. State Department to AT&T and the U.S. semiconductor industry, the TDMA folk had tied up all the politics and marketing. But the technicians in Europe's communications consortium in 1998 reversed their intense and tenacious hostility to CDMA. They concluded that no number of political endorsements or marketing programs or new physical laws enacted by the EEC could make TDMA work effectively for phones that combine bursts of data with streams of voice. The new version of GSM would have to be CDMA. Finally even Ericsson gave in. In early 1999, the company paid $4 billion to purchase Qualcomm's CDMA infrastructure manufacturing patents and facilities (which Qualcomm had created to demonstrate the technology). Like Intel, Qualcomm would focus its energies on designing chips and collecting royalties, while letting others build the systems.

The triumph of CDMA is an epochal event with implications that go beyond the prospects for Qualcomm. A new paradigm is moving to the fore. "Wide and weak," it uses bandwidth as a replacement for switching and power. This vision is at the heart of the most promising technologies of today, from advanced digital teleputers to ubiquitous mobile phones and computers in so-called personal communications services (PCSs). Wide and weak is even the secret of wavelength-division multiplexing (WDM) over fiber optics. Dispersion in optical fiber increases by the square of the power used to transmit, which tends to rise in step with the bit rate. The more bits you send per second on a single wavelength, the more power you use, resulting in a square law increase in dispersion that blurs the bits. This Catch-22 predicament Shannon surmounts also: Send lots of separate bit streams at lower power, as in WDM. Many low-powered bit streams down a fiber will outperform a single high-powered bit stream of time-slotted messages. Shannon's theories of the telecosm provide the basic science behind the optical miracles of WDM, the explosion of digital cellphones, and the huge promise of broadband wireless.

The move up the spectrum is accelerating. Personal communications services, the prevailing new form of cellular, operate at two gigahertz. MMDS from WorldCom and Sprint runs at 2.5 gigahertz and Tim Shepherd, a wireless enthusiast from MIT, projects that DSP and transmitter technology will soon be able to function in a vertiginous band between 59 gigahertz and 64 gigahertz that the FCC has been contemplating for unlicensed and experimental uses. Bob Metcalfe's venture find, Narad Networks, is contemplating transmission at these frequencies. Canon and Lucent have been testing campus-area wireless networks that send 155 megabits per second data over the same infrared carriers that dominate fiber optics. So has Jolt Technology of Israel. TeraBeam of Seattle is now popularizing wireless optics at gigabits per seconds. Shannon's laws suggest that the move toward wide and weak will impel vast increases in the cost-effectiveness of communications.

The great astronomer and physicist Kepler wrote: "I cherish more than anything else the Analogies, my most trustworthy masters. They know all the secrets of nature." Shannon's insight was based on an uncanny analogy of communications to multidimensional geometry. The center of the sphere is where power is unlimited and bandwidth scarce. The surface of the sphere is low powered but broadband. Shannon's analogy took the theory of the telecosm from the center of the sphere to its surface, where the results were weirdly wide and weak and counterintuitive. Such is the power of the telecosm.

Chapter 9

The Satellite Ethersphere

"They'll be crowding the skies!"

—Steven Dorfman

Thus spoke the president of telecommunications and space operations for GM Hughes—the colossus of the satellite industry—warning the world of a new peril in the skies. Planning to launch an announced 840 satellites (since reduced to 288 or fewer) in low-earth orbits, at an altitude of some 435 miles, were a gang of cellular phone jocks and computer hackers from Seattle going under the name of Teledesic. Led by Craig McCaw and Bill Gates, they were barging onto his turf and threatening to ruin the neighborhood.

You get the image of the heavens darkening and a new ice age looming as more and more of this low-orbit junk—including a total of some 700 satellites from Motorola's Iridium, Loral-Qualcomm's Globalstar, Loral-Alcatel's SkyBridge, and Teledesic, among other low-earth orbit (LEO) projects—accumulates in the skies. Ultimately, from this point of view, you might imagine the clutter of LEOs eclipsing the geostationary orbit itself, the so-called Clarke belt, some 21,000 miles farther out, where lots of satellites already float.

In an article in *Wireless* magazine in 1945, Clarke first predicted that satellites in orbit 22,282 miles (35,860 kilometers) above the equator, where the period of revolution is twenty-four hours, could maintain a constant elevation and angle from any point on earth. In such a fixed orbit, a device could remain for decades, receiving signals from a transmitter on the earth and radiating them back across continents.

The Clarke orbit posed some problems, however. One stems from the speed of light, which imposes a quarter-second delay on transmissions to geostationary satellites—enough to complicate or even prohibit interactive links such as voice conversations or Internet protocol exchanges. GEOs, as they are called, are not satisfactory vessels for phone calls and can only be linked to the net with special purpose protocols. The second problem is the inverse square law for signal power. Signals in space attenuate in proportion to the square of the distance they travel. This means that communications with satellites 22,000 miles away typically require large antenna dishes (as much as ten meters wide) or megawatts of focused beam power.

Now, however, a new satellite industry is emerging. It adapts to the light-speed limit through use of advanced microchips. These devices enable compact handsets with small smart antennas that can track low-earth orbit satellites sweeping across the skies at a speed of 25,000 kilometers an hour at a variety of altitudes between 500 and 1,400 kilometers above the earth. Twenty-five thousand kilometers an hour seems fast but it is radically slower than GEOs travel in their far larger orbits to keep up with the earth. Between twenty-five and sixty times nearer than geostationary satellites, LEOs find the inverse square law working in their favor, allowing them to offer far more capacity, cheaper and smaller antennas, or some combination of both. In their low and slow configuration, they also offer delays comparable to terrestrial fiber links. LEOs thus can be an integral part of the net. Using smaller orbits, these systems also vastly expand the total available room for space-based communications gear.

It is indeed technically possible to "crowd" the Clarke belt—a relatively narrow swath at a single altitude directly above the equator. But even this swath does not become physically congested; collisions are no problem. The Clarke belt becomes crowded because the ability of antennas on the ground to discriminate among satellites is limited by the size of the antenna. Teledesic plans to use what NASA calls the "Ka" band of frequencies, between 17 and 30 gigahertz. In this band, reasonably-sized antennas 66 centimeters (a little over two feet) wide can distinguish betweeen geostationary satellites two degrees apart. That's some 800 miles in the Clarke belt, thus no physical crowding. But it means that there are only a total of 180 Clarke slots for Ka band devices, including undesirable space over oceans.

LEOs, however, can be launched anywhere between the Earth's atmosphere and a layer of intense radiation called the Van Allen belt. The very concept of crowding becomes absurd in this 900-kilometer span of elevations for moving orbits that can be 500 meters apart or less. Thus the twenty-one proposed orbital planes of Teledesic occupy a total of 10 kilometers of altitude. At this rate, 225 or more Teledesic systems, comprising some 65,000 satellites, could comfortably fit in low-earth orbits.

Nonetheless, it was clear that the LEOs were crowding Hughes in a different sense. Hughes commands satellite systems or projects that compete with every one of the LEOs. Hughes responded to the threat of Teledesic by announcing the expansion of its Spaceway satellite system, then planned for North America alone, to cover the entire globe. Then, invoking the absolute priority officially granted geostationary systems, Hughes asked the Federal Communications Commission to block Teledesic entirely by assigning Spaceway the full five gigahertz of spectrum internationally available in the Ka band.

On May 27, 1995, Dorfman summoned the upstarts, Craig McCaw and Teledesic President Russell Daggatt, to Hughes headquarters in Los Angeles for a talk. Busy with Microsoft—which in 1993 surpassed the market value of Hughes parent General Motors—Teledesic partner Bill Gates did not make the trip. But as the epitome of the personal computer industry, his presence haunted the scene.

Together with Spaceway chief Kevin McGrath, Dorfman set out to convince the Seattle venturers to give up their foolhardy scheme and instead join with Hughes in the nine satellites of Spaceway. Not only could Spaceway's nine satellites cover the entire globe with the same services that Teledesic's 840 satellites would provide, Spaceway could be expanded incrementally as demand emerged. Just loft another Hughes satellite costing $150 million, with "every component proprietary to Hughes," as Dorfman boasted.

Spaceway's ultimate system, however, capitulated to the LEO logic. Announced in 1998 was a new Spaceway, deploying—along with eight GEOs—twenty NGSO (non-geostationary orbit) satellites, meaning LEOs. The new LEOs marked as decisive a break in the history of space-based communications as the PC represented in the history of computing. Teledesic would be the only LEO fully focused on serving computers with packetized data—the first truly "global Internet," as Teledesic board member Tom Alberg depicted it. It brings space communications at last into the age of ubiquitous microchip intelligence, and it brings the law of the microcosm into space communications.

Media observers, however, often did not like what they heard. Treating technology as a school for scandal, journalists offer mock-heroic exposés of computer hype, monopoly, vaporware, viruses, infoscams, net porn, securities "fraud," Y2K apocalypse, and deviously undocumented software calls. Dwelling endlessly on a "digital divide" yawning between the information rich and the information poor, pundits consign themselves undeniably, amid many yawns, to the latter category.

One such lugubrious saga began early in 1995 with McCaw and Gates, allegedly boarding McCaw's sleek yacht and going on an ego trip. With McCaw pitching in an early nickel, and the boat, and Gates hoisting his name as a sail, the two tycoons seemed to sweep away from the shores of rationality, as the media told it, into a sea of microwaves and arsenic (for millions of gallium arsenide microchips). Spinning out Teledesic to build an information superskyway, they came up with the seemingly cockeyed idea of launching 840 satellites, plus 84 spares. All would whirl around the world at a height of 700 kilometers (435 miles), using what they then told the FCC would be some 500 million gallium arsenide microchips. These would issue frequencies between 17 and 60 gigahertz from some 180,000 phased-array antennas serving both intersatellite links and connections between sky and earth. Following the 1997 announcement that Boeing would become an equity partner in Teledesic with a $100 million investment for a 10 percent stake, and become the prime contractor for the system, the constellation was redesigned as a 288-satellite system. Still, the entire project seemed suffused with gigahertz and gigabucks. "We're bandwidth bulls," says Teledesic President Daggatt.

In case the hype of the sponsors failed to keep the system radiant and aloft, fueling it also would be a total of 12,000 batteries fed by thin film solar collectors stretching out behind the satellite "birds" in some 130 square kilometers of gossamer wings. Working at 4 percent efficiency, these cells would collectively generate ten megawatts of power, enough to light a small city, but, so the critics said, insufficient to reach Seattle at microwave frequencies in the rain. (The Teledesic frequencies are readily absorbed by water in the air.) To manage the elaborate mesh of fast-packet communications among the satellites and to and from the ground terminals, the constellation would bear some 282,000 MIPS of radiation-hard microprocessors and a trillion bytes or so of rad-hard RAM. In effect, Teledesic would be launching into space one of the world's largest and most expensive massively parallel computer systems.

At a mere $9 billion, to be put up by interested investors, Teledesic's lawyers told the FCC, the price would be a bargain for the U.S. and the world. (By contrast, then current plans called for $15 billion just to lay fiber for interactive TV in California.) But former Motorola, then Kodak, chief George Fisher—fresh from pondering numbers for the apparently similar Iridium projects—suggested that $40 billion for Teledesic would be more like it.

Summing up a near-consensus of critics, John Pike, director of the Federation of American Scientists' Space Policy Project, declared to *The Wall Street Journal,* "God save us. It's the stupidest thing I've ever heard of!" Provoking Pike may have been the origins of the multisatellite architecture in the Star Wars "brilliant pebbles" program.

Already planned to be in place, however, were several other LEO projects, led by Motorola's Iridium and Loral-Qualcomm's Globalstar. As mobile phone projects, these systems could not readily offer service at T-1 data rates. But their sponsors promised availability for simple E-mail, faxes, and paging.

By mid-1994, Motorola seemed to command the financial momentum for Iridium. The company succeeded in raising some $800 million in equity investments from companies around the globe, including Lockheed and Raytheon (which would build the satellites), Great Wall of China and Khrunichev Enterprises of Russia (which together would launch a third of them), the Mawarid Group of Saudi Arabia (which pitched in $120 million), and Kyocera, Mitsui, and

DDI, which together put up another $120 million (Kyocera planned to build the dual-mode handsets for Japan and DDI to sell and service them).

Iridium's attractions were impressive. It promised ubiquitous global phone service at a premium price with little or no dependence on possibly unreliable local terrestrial facilities. But beyond the bold and ingenious concept, the system suffered from crippling technical flaws. As the project fell into bankruptcy in 1999, critics tended to ignore the technical flaws and cast aspersions on the very concept. It was said there was no market for this kind of service.

After all, irradiating much of the land mass of Europe, Asia, North America, and South America are terrestrial cellular antennas, LMDS (local multipoint distribution service) microwaves, and DBS (direct broadcast satellite) streams. In addition, coax and fiber lines and conduits loop redundantly around the globe. Every civilized point on the Earth pulsates with power cables and their hidden optic ground wires, and is thronged with trillions of meters of twisted-pair copper wires, their spectra bulging like engorged snakes with broadband signals using bulbous digital subscriber line (DSL) algorithms. In Bangladesh and Somalia, the Grameen Bank is making microloans of under $100 to village women to purchase chickens, cows, and cellular phones. Even in the most remote reaches of Kenya, devoid of electricity, so my daughter reports from the front, small huts fashioned with a fragrant blend of mud and dung are sprouting solar panels to power the TV. Cellphones will be next. Terrestrial wireless and wireline service is available everywhere, at a price far below the cost of satellite minutes. Isn't that why Iridium failed miserably and Globalstar and all other LEOs will fail as well?

What's wrong with this picture?

The key error is the assumption of widespread cellphone coverage. Even in the U.S., cellphones reach less than 20 percent of the territory. At least 50 percent of the country will never be economically served by cellular. In China, where there are five hundred thousand villages with no phone service at all, the current coverage is under 1 percent. Most of the world has no terrestrial phone service and will not have it for a decade or more. The opening price of Globalstar minutes was one seventh of Iridium's price and the Globalstar network remains nineteen times more cost effective. Inflation adjusted, one dollar per minute was the average price of long-distance telephony in the U.S. just twenty-five years ago. Incomes adjusted, the price of LEO service will be lower than most local telephony around the world just five years ago.

Doomed by catastrophically bad marketing by its regional telco partners, Globalstar was a key player in four dimensions of the paradigm—the collision with the light-speed limit, the primacy of dumb networks with all the intelligence on the edge, the superiority of Qualcomm's low-powered code division multiple access (CDMA), and the worldwide spread of wireless communications. Using CDMA and other architectural efficiencies, Globalstar escaped the functional rigidities and burdensome costs that blighted Iridium with its $7-per-minute long-distance charges. Globalstar began by charging thirty-five cents a minute for most wholesale links.

With just four hundred thousand customers, Globalstar could have broken even. The system can accommodate four million users. Adding twelve more satellites to the forty-eight bird constellation raises the total to six million. That means under $700 in total capital outlays per customer, less than the cost of laying a twisted-pair line to a home (without any other phone company facilities). A LEO customer could cost less than half as much money to service as a wireline phone customer paying less than one tenth as much per minute.

As the only way to reach the entire surface of the globe at once, satellites play a key role in the telecosm paradigm, but their advocates often misjudge their real promise. The currently dominant geosynchronous birds in the Clarke orbit will lose nearly all their network long-haul trade to fiber optics. At an estimated 8.6 terabits a second, a single fiber cable with 864 strands could hold more potential bandwidth than all the satellites, aloft or planned, put together. For point-to-point services, *satellites are relevant only where fiber does not reach.* Fortunately for the satellite industry, fiber reaches only a tiny portion of the world's destinations. Almost totally dominant for backbone services, fiber is sparse in the last mile. Thus rather than the backbone of the net itself, the most available prey for the tropospheric birds is links between customers and the net. Rapidly losing their remaining long-distance applications to fiber, satellites—though hundreds of miles above the surface of the earth—are emerging as major players in the "last mile."

The key innovator in LEOs was a Chinese engineer named Ming Louie. Although he was the best student in his high school, Louie was barred from advanced education during the 1960s cultural revolution because his father had gone to Columbia in the U.S. in the 1930s. After four years hauling logs in a saw mill, Louie had had enough by 1969. At age twenty-two, he procured an inflatable pillow, crept by night with a friend several hundred miles through the mountains to the Chinese coast, slathered his body with Vaseline for protection from the cold, and swam four hours across the straits to Hong Kong. There he got a job in a bakery (chosen because it had "plenty of food"). Happening into a bar during the Apollo lunar landing, he fixed on the image of the American flag planted on the moon. He decided to come to the U.S. and become a rocket scientist.

A determined and tenacious man, by 1989 he emerged as a rocket scientist working for Ford Aerospace in Palo Alto as a systems engineer on a project to connect all Ford automobiles to geosynchronous satellites for communications and locator services. He quickly encountered the light-speed limit that currently imposes the most critical constraint on all network design. Whether on the silicon substrate of a microchip, on the mostly silicon substrates of continents and sea beds, or between the surface of the earth and the usual satellite orbits, the velocity of light is a showstopper. Governing the latency of off-chip memories, turnaround times, and point-to-point links, light speed increasingly shapes the configuration of all systems. At Ford, Louie found it was simply impossible to overcome the crippling delays (a half second for two-way links) characteristic of the Clarke orbit twenty-three thousand miles away from the earth. With colleague Robert Wiedeman, he proposed a low-earth orbit system. Ford dismissed it out of hand. But when the company was sold to Loral, Louie and Wiedeman revived the idea. It became Globalstar.

The advantages of the Globalstar design stem only partly from its avoidance of complex intersatellite connections and use of infrastructure already in place on the ground. More important is its avoidance of exclusive spectrum assignments. Originating several years before spread-spectrum technology was thoroughly tested for cellular phones, Iridium employs the now obsolescent TDMA. In March 1990, Qualcomm CEO Irwin Jacobs, Allen Salmasi, the former Qualcomm executive who conceived Globalstar, and technical paragon Klein

Gilhousen all flew to Phoenix in a life-threatening storm in order to propose to Motorola that it combine with Qualcomm and use CDMA technology. But the Motorola officials refused. Fatefully they declared, "Iridium is cast in concrete."

Since 1994, I have predicted that "Iridium would sink like a stone." The Globalstar design can still lead the way to a new generation of two-way satellite services.

With their broad footprints, GEOs are intrinsically broadcast systems that can cover the entire world with downloads. But any attempt to provide millions of individuals with distinct Internet transmissions from the far corners of the web quickly exhausts transponder capacities and solar power budgets.

Among currently available consumer satellite Internet services, the standout is Hughes's DirectPC, which offers 400 kilobit downstream services and upstream links through telephone lines. But DirectPC is still not ready for prime time. The DirecTV–AOL alliance focuses not on the premium AOL Plus high-speed Internet service but on AOL TV available through DirecTV. AOL TV is a flakey interactive TV offering in the guise of Internet service. In addition to a program guide, AOL will offer such familiar interactive trivia as the lifetime batting average of the player at the plate, advertiser's product information, and other cosmetics for the corpse of broadcast television.

Similarly lame has been the EchoStar–Microsoft WebTV alliance. Purchased for $400 million and upgraded with estimated billions, WebTV has been a spectacular failure. It garnered just 800,000 subscribers while PC Internet access won scores of millions of users. The WebTV interface for EchoStar will allow an interactive search of 350 to 500 channels of video content and enable shows to be digitally recorded to a hard-drive for pause, rewind, and resume functions. But its Internet capabilities are meager and interactive TV will always pale in comparison to the near infinite choice available through the Internet.

A more promising broadcast application is caching the most popular web content for transmission over satellite links. But the amount and diversity of available content far exceeds what can be economically cached. Reported to catalog only 16 percent of web pages, the best Internet search engines are foundering under the load, taking as long as six months to add new content. Also swamped by a runaway web, caching systems face the same difficulty and dilemma as the search engines.

TCP puts an upper limit on the performance of a satellite. In TCP's congestion-avoidance mechanisms, when an acknowledgement fails to arrive within a certain time period, whether because of delays or errors, TCP reverts to a slower rate for retransmission. Based on research using NASA's ACTS (advanced communications technology satellite), Mark Allman, et al, in their 1997 paper, "TCP Performance over Satellite Links," report that the slow start mechanisms of TCP can require eleven roundtrip times of approximately 6.5 seconds, until a transmission reaches full speed.

While many of the problems can be mitigated by modifying TCP, the needed changes must often be made throughout the network. Between the satellite link and the Internet, this approach inserts a maze of gateways, proxy servers, and protocol translators that would add substantial costs and delays to the end-user Internet interaction. A simple law of the telecosm is: Don't bet against Internet standards in favor of tricky private solutions.

The only reason for tolerating the drawbacks of GEO latency is the lack of any alternative serving so many locations. As GEO users themselves increasingly recognize, LEOs are drastically changing that. In 1999, even Hughes finally gave up the ship, applying to the FCC for launch and operating authority for HughesNet, its own seventy-satellite LEO system at 1,490 kilometers.

LEOs fit the low-power paradigm. Signal strength drops by the square of the distance traveled, making low satellites far more power efficient than their GEO counterparts. Hundreds of LEOs each collecting enough solar energy for its individual low-power mission will be far more efficient than dedicated space-based power plants for GEOs.

Decentralization of service coverage also favors LEOs. GEO proponents have argued they can loft a couple satellites over densely populated areas and add service as demand grows elsewhere by adding new birds. Similarly General Atomics and others have proposed to launch dirigibles and circling automated airplanes to cover urban areas. But satellite service becomes paradoxically more valuable as populations thin out. It is economically perverse to build a market around the U.S. and Europe where communications capabilities and fiber networks are the best in the world.

For the LEO perspective, consider the plans of Teledesic, which will offer a 500 Mbps capacity in any circular area 200 kilometers in diameter. Such capacity in an urban area is almost laughable when cable, DSL, LMDS wireless, and direct fiber connections abound. But travel a short distance outside the city, head two hours north of New York City or west from Boston and you arrive in our own Berkshire Hills, a land that fiber forgot. We could use 500 Mbps. Head around the globe and you will find huge unsatisfied demand for cost-effective narrowband and broadband communications. But that demand is not centered on those who can afford $3,000 handsets offering voice at several dollars per minute, as Iridium learned.

Echoing Teledesic's promotion of LEOs for data was the Alcatel/Loral partnership SkyBridge, formed in June 1997. Unlike Globalstar with its focus on mobile voice and messaging, SkyBridge offers fixed broadband multimedia services through eighty satellites in two forty-satellite constellations at 1,469 kilometers. While featuring fewer satellites than Teledesic (288 satellites), the selection of CDMA for SkyBridge's transmissions should give the system an advantage in power and capacity compared to Teledesic's TDMA. SkyBridge hopes to provide more than 20 million users with download speeds of up to 20 Mbps and upstream links of 2 Mbps, with higher rates available through channel aggregation, another task easier using CDMA than with TDMA.

In plans for SkyBridge, Loral is also using the "bent pipe" strategy of Globalstar. Following the dumb network model of bits in, bits out, there will be no on-board processing of data. Data will take just one short hop from $700 end-user terminals to one of two hundred worldwide terrestrial gateways owned and operated by partnering regional service providers. Reducing the complexity of the in-sky network through a dumb-network strategy should improve the efficiency of the system, eliminate cross-network satellite-link delays, and offer the possibility of simpler, lower-cost satellites.

With SkyBridge's system cost estimated at just $6 billion compared to over $9 billion for Teledesic and service planned for the end of 2001, versus Teledesic's 2003, the cheaper, simpler strategy seems to be paying off. The concentration of intelligence in the ground segment also allows for greater flexibility in evolving the system to meet local service demands or future needs.

Dooming most GEO satellite approaches—from SkyCache to DBS's new "interactive" TV—is their assumption of a centralized, U.S.-centric, asymmetric network. These plans recall the old hope that a few generic TV networks providing least-common-denominator sitcoms could hold out against the variety of dozens, then hundreds, of cable channels. Now the GEO proponents hope that a few hundred or even thousands of interactive channels or gigabytes of cached pages can substitute for the diversity of millions of network users and the dynamic flexibility of millions of connected servers. But the Internet gains value at an exponential rate from the contributions of new users connecting to it, not because they are additional eyeballs for mega-site advertisers, but because each user can contribute to the content of the network.

Currently the two CDMA LEO projects Globalstar and SkyBridge are separate, one dedicated to narrowband mobile voice and the other to broadband fixed data. But both projects join Alcatel, Loral, and Qualcomm. With the data superiority of CDMA and all the intelligence on the ground, Globalstar's initially paltry 9.6 kbps messaging links could grow by the leaps and bounds of Moore's law to accommodate serious two-way data services in future generations. With no architectural changes, nine kilobits per second in 1999 means 64 kilobits per second in 2003, 128 kilobits in 2005 when SkyBridge will reach its prime, and a megabit per second in 2010. With larger solar panels and solar cell efficiencies, with better batteries, with enhanced modulation schemes, and with a new set of satellites slated for orbit in 2007, Globalstar should be scalable enough to force integration with SkyBridge. Indeed, the two CDMA systems can even jeopardize Teledesic if the Gates–McCaw project persists in its resolve to repeat the mistakes of Iridium.

So what is this, another saga of hubris on the "information superhighway"—to go with Al Gore's "invention of the Internet"? Perhaps good new ideas are harder to come by as company revenues grow into the billions, and Gates and McCaw disinvest and diversify as fast as they can from their increasingly cumbrous vessels of wealth. Having recently passed the billion-dollar mark in his systematic process of disinvestment from Microsoft, he retains $40 billion or so. Gates at times seemed embarrassed by his link to this gigantic project. He told us in 1995 it was too early to write about Teledesic.

Teledesic, however, is not merely a new global gambit by bored billionaires. It shares with Globalstar a new technological paradigm and takes it dramatically up spectrum. Early in the new century, the personal teleputer, summoning and shaping films and files of images from around the world, will collide with the centralized establishments of TV broadcasting. At the same time, Teledesic and the other LEOs will usher in the age of decentralization in space.

This change transforms the dimensions of the world as decisively as trains, planes, automobiles, phones, and TVs changed them in previous eras. It will extend "universal service" more dramatically than any new law can.

Moreover, Teledesic can eliminate the need to cross-subsidize rural customers. Determining the cost of wire-line services are the parameters of population density and distance from the central office. Rural customers now cost between ten and thirty times as much to serve with wires as urban customers do. Globalstar and Teledesic will bring near-broadband capabilities to everyone in the world at the same price.

Most important, this expansion of the communications frontier will foster the very economic development that will fuel the demand for the service. Today, it does not pay to bring telecommunications to poor countries that might benefit most. LEOs can break the bottleneck of development. Simultaneously opening the entire world, they enrich every nation with new capital exceeding the fruits of all the foreign aid programs of the era.

McCaw explains: "It'll be a disaster if China mimics what we did—building more and more urban towers and filling them up with people who queue up every day on turnpikes into the city, emitting fumes into the air, and then building new towers and new highways when you want to move the company, and then digging up the highways to install new wires."

McCaw waves toward the window, out at Lake Washington. "Look at that floating bridge. It took one and a half billion dollars to cross Lake Washington, then it got busted in a storm. Cross this lake, any lake, any ocean in the world with broadband wireless. That's the promise of Teledesic. All you do is to reconfigure the communications in software at zero incremental cost. No wires for the final connections. It's what we do in Hong Kong and Shanghai, where everyone uses a cellular phone."

Satellites can complement with ubiquitous access the global onrush of bandwidth dedicated to the Internet. As Alberg puts it, LEOs can give us "a truly global Internet in an ever-expanding ethersphere."

Chapter 10

The Coming of Component Software

"I was trying to conceive of how one could approach it in a way that would be fundamental . . . like being at the center of a sphere, where there were opportunities—and problems, of course!—in all directions."

—Patrick Haggerty of Texas Instruments, explaining his decision to license the transistor from Bell Laboratories in 1952

What will it take to launch a new Bill Gates—an Archimedean man who sharply shifts the center of the sphere, alters the axes of technology and economy, and builds a new business empire on new foundations? It is in part a symbolic issue, for there will never be another leader quite like this transcendental nerd tycoon. But the bandwidth explosion now overthrowing the calculus of abundances and scarcities of the world economy will also undermine Gates's supremacy.

The bandwidth tidal wave in hardware will necessarily impel a complementary change in software, and that's where the greatest economic power may reside. Truly abundant bandwidth will render your hard drive an inconvenient and expensive place to store the myriad software devices and features that you will want to use in the course of a year. Your own computer cannot even supply you with the various drivers and translators that you need in the course of a browse on the net. When computer users typically spend more time tapping remote disks and databases than accessing their own, software will be fundamentally changed. It will move out of Microsoft's favored channels and form factors and proliferate across the web in components rather than in suites. As this trend plays out, the pivotal question remains in force: Who can inherit the imperial throne in the microcosm and telecosm currently held by the Redmond Rockefeller?

I will open the envelope in a minute. But I should warn you that things have changed a bit since August 1995 when I first offered this promise. Some of the names have changed, but the key developments remain essentially the same.

The new era in software began with a software program called Netscape Navigator Personal Edition. I brought it back from Silicon Valley in late June 1995 and put the package next to my PC. The PC was proudly running a beta version of Windows 95. I had presented Windows 95 with great fanfare to my then eleven-year-old son Richard as his route to the most thrilling new frontiers of the computer world. Multitasking, thirty-two-bit operation, flat memory! Object linking and embedding! "Information at your fingertips!" You remember these incantations of the day.

I live out in the boondocks of western Massachusetts where there were then no convenient full-service connections to the Internet. So I was initially much less excited about Netscape than I was about Windows 95. I hoped Windows 95 would put me online through the Microsoft Network system. Some ten minutes later, though, Richard wanted to know my credit card number so I could choose an Internet service provider. A couple of minutes after that, linked through Internet MCI's 800 number, Richard was on the World Wide Web, using the InfoSeek service to examine my chapters from Telecosm online, searching the secrets of Sim City 2000 at Maxis, exchanging messages with Microsoft Flight Simulator buffs, and exploring Disney. There he remained pretty much full time for the next four years. He started a company to sell airline books. Contrary to the headlined social science of 2000, he did not become discernibly antisocial.

The next thing I knew, my brother Walter came by. Now at an ISP named Ziplink, he then worked for a computer company, New World

Technologies in Ashland, Massachusetts, that built customized Pentium machines and delivered them to value-added resellers within forty-eight hours. Walter wanted the Netscape program. He took it back to my parents' farm down the road and booted it up on a four-megabyte outdated PC previously used to map the pedigrees of a flock of mongrel Romney sheep. Soon he was on the web scouting out the competition from Dell and Micron and showing off the Gilder web page. This intrigued my then seventy-seven-year-old mother, who had scarcely even noticed a computer before. I don't know how it happened, but before the night was out, she too was on the web, exploring catalogs of British colleges for her namesake granddaughter who was soon to leave for London.

Remember that time? The initial exposure to the web through Netscape was the nerd's version of first love.

Now let me tell you about my introduction, also in 1995, to the Java programming language and virtual machine. Java was the first step toward a new world of software programming that menaced Microsoft's supremacy. I encountered Java in early June 1995 at a Sun Microsystems conference at the Westin St. Francis Hotel in downtown San Francisco. For a speech I was to give, I had planned to use a multimedia presentation, complete with Macro-Mind Director images and QuickTime video that I had contrived with an expensive professional some months earlier. The complexities of Director prompted me to convert the program to an allegedly simpler presentation program called Astound. However, it required an external disk drive and ran erratically with the eight megabytes of RAM on my PowerBook. I decided to speak nakedly from notes on the coming technologies of sand and glass and air.

Following me immediately to the stage was Sun's amiable chief scientist, John Gage. He decided to illustrate his speech entirely from the World Wide Web. He began with a handsome page, contrived minutes before, giving an account of my speech, headlined: "Gilder Addresses Sun, Tells of Technologies of Opaque Silicon and Transparent Silicon." Then he moved to the Gilder Telecosm archives run by Gordon Jacobson of Portman Communications at a web site of the University of Pennsylvania's engineering school. Gage illustrated his talk with real-time reports on traffic conditions in San Diego (where I was about to go), weather conditions in Florida as a hurricane loomed, and developments on Wall Street as IBM bid for Lotus. He showed the Nasdaq ticker running across the screen. He showed animations of relevant charts, cute little Java gymnasts cartwheeling across the screen, three-dimensional interactive molecular models, and an overflowing coffee cup, entitled "HotJava."

None of his information and images used a desktop presentation program. None of them used a database engine. Indeed, except for the Gilder speech report, none were created beforehand. Incurring no memory or disk drive problems, Gage summoned all the illustrations directly from the Internet. The animations employed Java, the then new computer language written for the web by the venerable Sun programmer James Gosling. Java allows transmission of executable programs to any computer connected to the net to be interpreted and played safely and securely in real time. It remains the most powerful promise in computing.

Clifford Stoll, the new wave cryptologer, calls it "silicon snake oil." But I call it a fundamental break in the history of technology. It is the software complement of the hollowing out of the computer by the emergence of bandwidth abundance. Almost overnight, the CPU and its software have become peripheral; the network, central. I had spent days working on a presentation on my desktop computer, using an array of presentation software. But Gage improvised a more impressive and animated presentation without using any desktop presentation programs at all. The World Wide Web and the Java language were enough. Restricted to the files of my computer, I struggled with storage problems and incompatible research formats, while he could tap the storage capacity and information resources of the then more than five (now sixty) million host computers on the net.

Similarly for my family, the limitations of my parents' barnyard PC didn't matter. The operating system also didn't matter. What was crucial was the network gear and software. My brother Walter had installed a modem that linked to the web at an average of 24.6 kilobits per second. With the Netscape Navigator, that was enough. Actually 14.4 would have been enough. Enough to launch a valid pretender for the role of a new Bill Gates.

Admit it, the legacy version, once so luminous, is beginning to lose its shine. You thrill no longer at his vaunted $50 million underground mansion, fenestrated with $40 billion Windows, offering misty views of Daytona, Memphis, Cairo, and other far-off places you no longer really care to go, even if they are swimming with screensavers from London's National Gallery.

At the most essential level, Bill Joy of Sun illuminated Gates's dilemma at Esther Dyson's PC Forum conference in 1990. Known as one of the great minds in software, yet losing share inexorably to Microsoft, Joy seemed to be moving into the role of conference crank. Year after year, he lamented the prolix inelegance of the triumphant waves of Microsoft programs sweeping through the industry: "As we add more and more of these features to older systems," he said, "the complexity gets multiplicative. I have ten different packages that interact in ten-to-the-tenth different ways. I get all sorts of surprises, and yet because these things don't play together well the power is only additive. I get this feature and that feature but the combinations don't work. What I'd really like to see is a system where the complexity goes up in a linear way but the power goes up exponentially."

In software, complexity has long been rising exponentially, while power has been rising additively. In response, Niklaus Wirth, the inventor of Pascal and other programming languages, has propounded two new Parkinson's laws for software: "Software expands to fill the available memory" and "Software is getting slower more rapidly than hardware gets faster." Indeed, newer programs seem to run more slowly on most systems than their previous releases.

But none of this matters. Gates has moved from triumph to triumph by shrewdly exploiting the advances of microcosmic hardware. With Moore's law and the Law of the Microcosm, the number of transistors on a chip doubles every eighteen months, and cost/performance rises as the square of the number of transistors. The complexity sinks into the microcosm and power rises exponentially on the chip to absorb all the grenades rolling down from Redmond.

Gates travels in the slipstream behind Moore's law, following a key rule of the microcosm: Waste transistors. As Nicholas Negroponte puts it, "Every time Andy [Grove] makes a faster chip, Bill uses all of it." Wasting transistors is the law of thrift in the microcosm, and Gates has been its most brilliant and resourceful exponent.

Meanwhile, in the face of Gates's ascendancy, Bill Joy seemed to grow curls and shed influence, as Sun played rope-a-dope with

Hewlett-Packard and other workstation rivals. In 1990, he retreated to a sylvan aerie in Aspen, Colorado, to pursue "advanced research" for Sun. But his talk of small programs and handheld consumer appliances seemed irrelevant to the company.

Nonetheless, in early 1990, showing up late—in a Hawaiian shirt—to address a formal dinner in Silicon Valley, Bill Joy had a great prophetic moment. He cited the Moore's law trends as grounds for granting the first five years of the 1990s to Bill Gates. "It's pretty much determined," he said. Indeed, in the late 1980s, Joy had personally made a separate peace with Microsoft by selling a large portion of his Sun holdings and buying Microsoft shares, thus becoming the second richest of Sun's four founders. (The richest, Andy Bechtolsheim, jumped even deeper into Microsoft.) But then, around 1995, predicted Joy, everything would change. There would be a "breakthrough that we cannot imagine today." He even acknowledged that the breakthrough would not come from Sun, but "from people and companies we cannot know today."

The key to software innovation, he said, was smart programmers. Smart programmers are hundreds of times more productive than ordinary programmers. And "let's be truthful," said the sage of Sun, propounding what has become known as Joy's law, "most of the bright people don't work for you—no matter who you are. You need a strategy that allows for innovation occurring elsewhere." To the Justice Department, Microsoft's overwhelming OS market share and its teeming armies of programmers seem a barrier to entry for other software competitors. To Joy, Microsoft's size and dominance could become a barrier to entry for Microsoft, blocking it from the key new markets of the late 1990s.

It is now clear that Joy was on target. The breakthrough is here in force, invading and occupying all the commanding heights of the information economy, from the media to the universities. It is the World Wide Web and its powerful browsers, servers, languages, and programming tools. Software on individual machines still bogs down in the macrocosmic swamps of complexity. But in the telecosm, yields rise exponentially almost without limit in proportion to the number and power of the machines on the network.

For the last five years, the traffic on the network has been rising between five and ten times faster than the number of transistors on a chip. With 4,300 miles of fiber-optic lines being laid every day in the U.S., bandwidth is sure to rise even faster than the number of networked computers. This awesome transition presents a supreme chance for new leadership in developing software focused less on wasting transistors than on wasting bandwidth.

A computer on every desktop and in every home? Information at your fingertips? *The Road Ahead?* My son Richard yawns. Let's face it, Bill, that stuff is yesterday. In the new era, Microsoft can continue to feed on the microcosm. But the leading-edge companies will move to the frontiers of the Telecosm, where collectively they will grow far faster than Microsoft.

So, lets open the envelope for candidate number one for Bill Gates' throne.

Start by adding one hundred pounds of extra heft, half a foot of height, and two further years of schooling, then make him $30 billion hungrier. Give him a gargantuan appetite for pizza and Oreos, Bach, newsprint, algorithms, ideas, John Barth, Nabokov, images, Unix code, bandwidth. Give him a nearly unspellable Scandinavian name—Marc Andreessen.

Put him to work for $6.85 per hour at Illinois' National Center for Supercomputing Applications (NCSA), writing 3-D visualization code. Surround him on all sides by the most advanced computers and software in the world, under the leadership of cybernetic visionary Larry Smarr. What will happen next? "Boredom," Andreessen replies. Supercomputers, already at the end of their tether, turned out to be "underwhelming Unix machines."

Then, for a further image of the end of the world, take him in the fall of 1990 off to Austin, Texas, for two semesters at IBM. "They were going to take over the 3-D graphics market, they were going to win the Malcolm Baldridge Award, they were going to blow Silicon Graphics [the regnant Silicon Valley 3-D workstation company] off the map, all in six months." Andreessen began by doing performance analysis and moved on to work on the operating system. In mid-1991, after constant delays, the company was finally ready to ship a world-beating 3-D engine. But the new IBM machine turned out to be four times slower at seven times the price of the equivalent Silicon Graphics hardware that IBM had distributed a year and a half earlier with its workstations.

Austin IBM returned to the drawing board and Andreessen returned to Illinois to get his degree.

In both commercial and academic settings, Andreessen thus had the good fortune of working at the very heart of the old order of computing in its climactic phase. As Andreessen saw it, little of long-term interest was going on at either establishment. But both did command one huge and felicitous resource, vastly underused, and that was the Internet. "Designed for all the wrong reasons—to link some two thousand scientists to a tiny number of supercomputers," it had exploded into a global ganglion thronged by millions of people and machines.

Many people saw the Internet as throbbing with hype and seething with problems—Clifford Stoll's *Silicon Snake Oil* would later catalog many: the lack of security, substance, reliability, bandwidth, easy access; the presence of porn, fraud, frivolity, and freaks. But to Andreessen the problems of the Internet were only the other side of its incredible virtues.

"By usual standards," said Andreessen, "the Internet was far from perfect. But the Internet finds its own perfection—in the millions of people that are able to use it and the hundreds of thousands who can provide services for it." To Andreessen, all the problems signaled that he was at the center of the sphere, gazing in wild surmise at "a giant hole in the middle of the world"—the supreme opportunity of the age.

Andreessen saw that, for all its potential, there was a monstrous incongruity at the heart of the Internet. Its access software was at least ten years behind. "PC Windows had penetrated all the desktops, the Mac was a huge success, and point-and-click interfaces had become part of everyday life. But to use the net you still had to understand Unix. You had to type FTP [file transfer protocol] commands by hand and you had to be able to do address mapping in your head between IP addresses and host names and you had to know where all the FTP archives were, you had to understand IRC [Internet relay chat] protocols, you had to know how to use this particular news reader and that particular Unix shell prompt, and you pretty much had to know Unix itself to get anything done. And the current users had little interest in making it easier. In fact, there was a definite element of not wanting to make it easier, of actually wanting to keep the riffraff out."

The almost miraculous key to opening up the Internet was the concept of hypertext, invented by Theodor Holm Nelson, the famously fractious prophet of the Xanadu network, and son of Celeste Holm, the actress. A hypnotic speaker, with a gaunt countenance and flowing golden hair, Nelson seems an Old Testament Jeremiah from central casting as he rails against the flaws and foibles of current-day computing. Hypertext is simply text embedded with pointers to other text, instantly available by a point and click. For the source of the concept, Nelson quotes an essay by Vannevar Bush written in 1945 and read to him by his father as a boy: "The human mind . . . operates by association. With one item in its grasp, it snaps instantly to the next that is suggested by the association of thoughts, in accordance with some intricate web of trails carried by the cells of the brain." Projecting this idea from a single human brain to a global ganglion, Nelson sowed the conceptual seeds of the World Wide Web. He imagined a network that would allow users point-and-click navigation wherever they wanted to go.

Andreessen can explain both the power of hypertext and its slow emergence in commercial products: "Xanadu was just a tremendous idea. But hypertext depends on the network. If the network is there, hypertext is incredibly useful. It is the key mechanism. But if the network is not there, hypertext does not give you any of the richness." Hence, Apple's HyperCard and similar schemes failed to ignite. The link is not hyper if it is restricted to your hard drive or CD-ROM. Connected to millions of computers around the globe, it becomes exponentially hyper.

Nelson's idea led to what Gary Wolf, a contributing writer of *Wired,* calls "one of the most powerful designs of the twentieth century"—a universal library, a global information index, and a computerized royalty system. But Nelson's quest for perfection led to a twenty-year adventure in futility. "The opinion of the Xanadu people to this day is that the web and the Internet are much too simple. They don't solve the problems. For instance, the links aren't fully bidirectional. You don't know exactly who's pointing to your page, and there's two ways to look at that. The way that Ted Nelson looks at it is, 'That's bad!' The way I look at it is, 'That's great!' All of a sudden anyone can point to your page without permission. The net can grow at its own rate. You get the network effect, you get Metcalfe's law, it spirals completely out of control. Isn't that fantastic?"

In 1988, Xanadu found funding from John Walker, the charismatic recluse, taxpatriate in Switzerland, and founder of Autodesk, the desktop computer-aided design company. As Walker prophetically declared in 1988: "In 1964, Xanadu was a dream in a single mind. In 1980, it was the shared goal of a small group of brilliant technologists. By 1989, it will be a product. And by 1995, it will begin to change the world." All truer than Walker could have imagined, but it would not happen, alas, at Autodesk. Haunted by dreams of perfection and hobbled by hyperventilation at the helm, Xanadu misted over. Autodesk sustained the effort until it was mercifully terminated by the new CEO, Carol Bartz, in 1993 at the very moment that a real Xanadu, deemed hopelessly imperfect and inadequate by Nelson, was about to burst forth efflorescently on the Internet.

The demiurgic step, now an oft-told tale, came from Tim Berners-Lee at CERN in Switzerland, creating the World Wide Web based on a universal hypertext function. He launched initial HTTP (hypertext transfer protocol) governing transport on the Web. He developed uniform resource locators (URLs) as a common addressing system that joined most of the existing Internet search and linkage technologies. He conceived the HTML (hypertext markup language), a kind of PostScript for the web. He made the system ignore failures (Joe's moving data could not crash your machine). Thus, the user could point and click to information anywhere on the net, unconscious of whether it was in the form of a file transfer, an E-mail, a Gopher search, or a news posting, or whether it was in the next room or in Tasmania.

Berners-Lee addressed a basic problem of the Internet from the point of view of an academic researcher. But the real opportunity was to open the Internet to the world and the world to the Internet, and that would require more than a facility for cruising through textual materials. After all, the bulk of human bandwidth is in a person's eyes and ears. And human bandwidth is quite slow: For absorbing text, as Robert Lucky, author of *Silicon Dreams,* has pointed out, the speed limit is only some 55 bytes per second.

To burst open the Internet would require reaching out to the riffraff who travel through pictures and sounds at megahertz speeds. At CERN, Berners-Lee opposed images and video on these grounds. The technologists all held a narrowband view of the world, imagining bandwidth as an essentially scarce resource to be carefully husbanded by responsible citizens of the cybersphere.

So Tim Berners-Lee alone could not burst open the Internet piñata and give it to the world. It was the ultimate broadband booster, Marc Andreessen, working with NCSA colleague Eric Bina, who ignited the web rocket. One late December night in 1992 at the Espresso Royale cafe in Champaign-Urbana, Andreessen looked his friend in the eye and said: "Let's go for it."

Every Gates has to have his Paul Allen (or Jobs, his Steve Wozniak). Andreessen's was Bina—short and wiry where Andreessen is ursine, cautious where he is cosmic, focused where he is expansive, apprehensive where he is evangelical, bit-wise where he is prodigal with bandwidth, ready to stay home and write the code where Andreessen is moving on to conquer the globe. Wildly contrasting but completely trusting and complementary, these two—in an inspired siege of marathon code-wreaking between January and March 1993—made Mosaic (the first visual browser) happen. A rich image-based program for accessing the World Wide Web and other parts of the Internet, Mosaic required no more knowledge of its internal mechanics than is needed by the user of the steering wheel of a car. With a mere 9,000 lines of code (compared to Windows 95's 11 million lines, including 3 million lines of MSN code), Mosaic would become the most rapidly propagated software program ever written.

Andreessen could defy all the fears of an Internet image crash because he lived in a world of bandwidth abundance and fiber galore. He fully grasped the law of the telecosm. Every new host computer added to the net would not only use the net; it would also be a new resource for it, providing a new route for the bits and new room to store them. Every new flood of megabyte bit maps would make the net more interesting, useful, and attractive, and increase the pressure for backbones running at gigabits per second and above. The Internet must be adapted to people with eyes and ears. They won't abuse it, he assured Bina without a smile. After all, he knew he would have to rely on Bina for much of the graphics coding.

"I was right," Bina says now. "People abused it horribly. People would scan in a page of PostScript text in a bit map, taking over a megabyte to display a page that would take maybe one thousand bytes of text. But Marc was also right. As a result of the glitz and glitter, thousands of people wasted time to put in pretty pictures and valuable information on the web, and millions of people use it."

Working night and day at the NCSA, wrangling over issues, arts and letters, music, and Unix code at the Espresso Royale down the street, the two programmers achieved a rare synergy. "We each did the job that most appealed to us," says Bina, "so each of us thinks the other did the hard stuff." Bina wrote most of the new code—in particular, the graphics, incorporating color management tools. Andreessen took the existing library of communications code from CERN and tore it apart and rewrote it so it would run more quickly and efficiently on the network.

As time passed, they brought in other young programmers from the NCSA—among them Chris Wilson, Jon Mittelhauser, Chris Houck, and Aleks Totic—to port the system to Windows and Mac machines. With help from them, they designed dynamic forms with type-in fields, check boxes, toggle buttons, and other ordered ways of entering simple text for searches and other functions. (This is the origin, for example, of the dynamic form now used by Pizza Hut for ordering a pizza online.)

In the end, they had created an entirely new interface for the Internet and new communications software to render it crisply accessible—a look and feel that almost immediately struck everyone who used it as an amazing breakthrough. In February 1995, Bob Metcalfe wrote a prophetic column in *InfoWorld* predicting that web browsers would become, in effect, the dominant operating system for the next era.

That is what happens when an entrepreneur performs a truly revolutionary act, supplies the smallest missing factor, as Peter Drucker puts it, that can transform a jumble of elements into a working system—the minimal mutation that provokes a new paradigm. In 1977, the relevant jumble was small computers, microprocessors, and assembly language programming. Bill Gates and Paul Allen supplied the key increment: software tools and the Basic language for the embryonic personal computer. In 1993, Andreessen and Bina set out to supply the minimal increment to convert the entire net, with its then one-to-two million linked computers (today it's an estimated 200 million computers) and immense information resources, into a domain as readily accessible to an eleven-year-old as a hard drive or CD-ROM on a Mac or Windows PC.

After Andreessen's college graduation in December 1993, Joseph Hardin, head of the NCSA, asked him to stay on at the center. But he added one condition, namely, that the young programmer leave the Mosaic project. "Some forty people had a role in creating Mosaic," he said. "Don't you think it's time to give someone else a chance to share the glory?" Reasoning that they would have a still better chance in his absence, Andreessen left for Silicon Valley.

After a few months doing Internet security functions at a company called EIT, Andreessen attracted the attention of the legendary Jim Clark, founder of Silicon Graphics Inc. and inventor of the Geometry Engine, capable of 3-D graphics, that impelled SGI's growth. At loose ends, like most of the industry's venturers, Clark was looking for opportunity in all the wrong places—3-D games, interactive TV, Hollywood. He found himself increasingly entranced with the Mosaic browser. Hearing that the author of Mosaic had arrived in Silicon Valley, he sent Andreessen an E-mail in early February 1994.

The rest is history, in the accelerated form familiar in the Valley. Jim Clark met with Andreessen and signed him up as the first employee of a new company. "What has happened to the other NCSA programmers?" he wanted to know. Hearing that Chris Wilson had left the center to join SPRY, Clark decided to act fast. He invited Bina to come to Silicon Valley and meet with him on his yacht. Avoiding temptation (Bina's wife was a tenured professor of database technology at Champaign-Urbana and is wary of snake oil), Bina spurned the invitation. Andreessen wrote him a glum E-mail: "Sorry. It would have been nice to have had you here."

The next thing Bina knew, Andreessen called him to announce that he and Clark were flying to Champaign-Urbana and wanted to meet with Bina and the rest of the key programmers the next day at the local University Inn. "When I got off the elevator at the hotel," Bina recalls, "this blond guy who reminded me of my dad stepped back, looked me slowly up and down, and spoke: 'Marc said you could walk on water. I have never seen anyone who could do that before.'" Clark soon assured him he could stay in Illinois.

According to Bina, Clark spread a contagion of entrepreneurial excitement. Then one by one he made seven of them offers. All signed up. By the first week of April, Mosaic's writers were on the way and Jim Clark forgot about interactive TV.

Greeting this bold new company, however, were rumblings about a possible intellectual property suit from the NCSA. This surprise banished any idea of using Mosaic in any form. That was fine with them. They knew Mosaic was a quickly written hack designed for high-speed college-type connections. "We knew that everyone had Mosaic," says Bina. "We were glad to start from scratch again." Moving into a small office in Mountain View at a cost of less than $1 per square foot, they set to work.

In the new company, then called Mosaic Communications, Andreessen ascended to management. Bina, working in Illinois for all but one frenzied week a month, immersed himself in creating the new code. They revised it to work over a 14.4 modem. They focused on making it the only system that is fully secure. They added new supports for more elegant layouts and richer documents. Giving the program away on the net, within a few months Netscape won 70 percent of the web browser market, and it retains a good chunk of it despite "monopolistic" competition by Microsoft.

As Wall Street began pouring money on any rival company with an Internet product, the stock market became a Netscape imperative. This posed a problem from Clark. There would be no difficulty attracting a frenzy of interest. The question was, Where would they hide Andreessen during the road show? Introducing him to John Doerr of Kleiner, Perkins, Caufield & Byers and other Silicon Valley clout, however, Clark instead resolved to teach the young bear how to chew with his mouth closed, tie a tie, and get to work earlier than 8:30 . . . P.M. Then he prepared to place him in the nose cone of an Initial Public Orbiting that would value the gang at several billion dollars, point Andreessen toward the future and, with luck and the right characterological chemistry, have a new Bill Gates.

Here he is, well after 8:00 P.M., roaring down El Camino Real in his '94 red Mustang, filled with pizza rinds, empty cans of Sprite, a mostly shredded copy of the *San Jose-Mercury News* and a bulging issue of *Micro Times* smeared in popcorn butter, talking with both hands to a writer from *Forbes ASAP,* hunched amid the clutter and stealthily searching for a seat belt.

Amazingly free of the NIH (not invented here) syndrome afflicting many of his elders—his browser combines products from some seventeen sources—Andreessen was discoursing on the virtues of Java, a language he says portends a telecosmic advance even more profound and enduring than the browser and related servers themselves.

For creating innovative stuff on the web, Andreessen explains, "The alternative languages are grim. None of them are truly suited to the web environment. Java is as revolutionary as the web itself." Tying Netscape to Java assures the company a central role in a coming efflorescence of the web more significant in its implications than even the current manifestation. In the process, it will invert the entire world of software in a way that may permanently displace Microsoft from the center of the sphere.

Behind this development is not some twenty-four-year-old geek from a supercomputer center, but the industry's most venerable leader after Bill Gates, not only Microsoft's most persistent critic and nemesis, but also its most eminent defeated rival: Bill Joy of Sun. In engineering Mosaic and Netscape Navigator, streamlining the communications functions and putting it all on the net, Andreessen was following in the codesteps of Joy. And just to hedge my bets a bit, I have to include the great Java maker as another possible Bill Gates, because you never know.

Joy was not only a founder of Sun Microsystems, he was also the primary champion of the Berkeley brand of Unix that he husbanded as a student at the university. So-called BSD Unix was not only Sun's essential software technology, but also the reference code for TCP/IP (Transmission Control Protocol/Internet Protocol), the basic protocol stack of the ARPAnet in its evolution toward the current Internet. Thus Joy has had a hand in several key pieces of networking software.

Indeed, because Sun computers running BSD Unix were the most robust vessels for TCP/IP, the lion's share of all host computers on the web are still Sun's, and Sun is experiencing an unexpected surge of earnings in the wake of the web. Now the computer world is wielding Marc Andreessen's browser to rush toward the net, wrenching the sphere from its moorings in Redmond, and allowing it to oscillate toward a new center in Silicon Valley. Even Bill Gates, the last holdout against Joy's network standard, bundled TCP/IP with Windows 95. Joy must feel that the industry is coming home at last.

Just as Gates ignited the personal computer market by writing Basic for the PC, Joy is launching a new era on the net by supplying a suitable software language, Java. Joy foreshadowed Java at the 1990

PC Forum conference when he said, "Large programs are embarrassing because they have a fixed set of ideas and so much code that it's very difficult to change them. And they all tend to reflect existing metaphors, not the new metaphors. I'm starting a small group to try to do something small—Sun Aspen Smallworks. I believe it's possible to do small systems of a few hundred thousand lines of code that live in this world of persistent distributed objects with open protocols [and] make an incredible difference—much more than an extra hundred thousand lines on a ten-million-line system. So I'm looking for a few great hackers."

The one he found was redoubtable Sun engineer James Gosling. Considered by John Doerr of Kleiner Perkins to be "perhaps the world's greatest living programmer," Gosling is a bearded man with shaggy blondish hair. He displays a subtly wounded look that is possibly the effect of a career that, to that point, had consisted mostly of brilliant failures. But, like Joy, Gosling was eager to do something small to bring Sun into the world of consumer electronics where a large program used no more than a few thousand lines. In 1992, the company spun out a subsidiary named FirstPerson Inc. to pursue this market.

The goal was to conceive a way to take news, images, animations, and other real-time functions and download them to handheld devices such as digital cellular phones. But with the Newton fiasco, personal digital assistants began to fall from favor and Sun and the rest of the lemmings decided to pursue the suddenly seductive siren of interactive TV. Again, they would use the Java (at the time called "Oak") programming language, this time running not on cellular phones, but in set-top boxes. Here, too, the programs would have to be elegant and compact to use the sparse memories affordable for a consumer appliance. However, like Jim Clark—and everyone else—Sun would soon discover that interactivity merely offers cosmetics for the corpse of a dying television industry.

At this point, the project was in turmoil and former Apple designer Wayne Rosing, its first manager, was on his way out. In May of 1993, Sun technical chief Eric Schmidt entered the breach, collapsed the set-top project into a joint venture with Thomson Consumer Electronics, which had bought RCA, and summoned Joy from Aspen. Joy returned and spent the entire summer in Palo Alto working with Schmidt and

Gosling to figure out a strategy for Java. As the work proceeded, it became clear that the characteristics of a language suitable for PDAs not only gratified Joy's laws of software, but also fit almost perfectly the needs of the web.

Like many technology projects that seem to spring full-blown from the brow of a genius, Java is in fact the fruit of a near-lifelong quest. Gosling's career began at age fourteen in Calgary when, during a high school tour, he memorized the code to the locks on the doors of the computer center at the University of Calgary. He then regularly admitted himself to read computer texts and use the available machines. He became so proficient that he was hired a year later by Digital Equipment to write machine code to be used for analyses of data on the aurora borealis from Isis satellites.

Going on to Carnegie Mellon University to study under the eminent Robert Sproull and Raj Reddy, he wrote a text editor called EMACS, a Pascal compiler, and a thesis entailing the creation of a program to do drawings of levers and strings in visual, interactive models. After graduation in 1981, he joined an IBM lab in Pittsburgh, where he developed the Andrew Windows system. The first major program that could control a window on a computer remotely across a network, it never found a home. But it launched Gosling on a fifteen-year struggle with the challenge of how to send programs across a network that could be executed on other computers.

Gosling had a breakthrough in 1983 at a Sun Microsystems conference at the Red Lion in San Jose. There, Joy observed Gosling's Andrew Windows system dramatically outperforming a workstation running Sun View, and began a concerted effort to recruit him. Soon after, Gosling came to Sun to develop a remote windows system called NeWS (networked extensible windows system). Initially received with great critical acclaim in the industry, it was eventually blocked by the X-Windows consortium despite performing far more efficiently over a network.

Then came Sun's ill-fated pursuit of consumer electronic products. Gosling set out to develop appropriate programs in C++. But it soon became clear to him that this language failed on nearly all the crucial criteria. In particular it was neither reliable nor secure, which, funnily enough, turned out to be two sides of the same coin. "The gun you shoot the burglar with can also shoot you in the foot." That was C++.

So Gosling began revising C++ code into something that has been termed "C++ minus minus." Returning to the inspiration of Andrew and NeWS, he adapted the language to enable real-time operations across a network. In order to run a variety of different programs on a variety of linked devices, from set-top boxes to palmtop remotes, he made it inherently platform-neutral and nomadic. It would have to be a language at home on a network.

The project ended with Java, an efficient programming language that is safe, simple, reliable, and real-time, yet familiar to anyone who has used C or C++. Designed for PDAs on infrared networks, it turned out to be perfectly adapted for the Internet. Many of Java's virtues are made possible by the fact that it is interpreted rather than compiled—that means that it is translated line by line in real time in the user's computer. It can be run by anything, anywhere, instantly. This makes all the difference.

It means that programs no longer have to reside in the machine where they are used, or be written for that machine in order to be executed by it. Potentially, a Java program can reside anywhere on the entire Internet and be executed by any computer attached to the Internet. All that's needed is a relatively tiny "interpreter" program, which most people get automatically, never think about, and never worry about. The little interpreters make Java programs "dynamically portable" in real time.

Dynamically portable programs are suitable for a nomadic existence on the net, rather than a mere settled life on the desktop. Java thus emancipates software from computer architecture. It offers a software paradigm radically different from the Microsoft model, which is based not only on static compilation, but on often-concealed, proprietary source code.

Suddenly, the entire world of new software is potentially available to every computer owner. Rather than being restricted to the set of programs you own, you can use any program on the net, just as now you can tap any information on the net. You not only have data at your fingertips, you have programs at your fingertips. This means "executable content," as Joy describes it and as John Gage first demonstrated in his speech at the Westin St. Francis Hotel in June 1995. Whether a film, a graph, an animation, a real-time bit stream on the Nasdaq ticker or the Reuters wire, a virtual reality visualization, or a game, it can be downloaded to your machine with its program in tow.

Owning the operating system and associated tool libraries becomes irrelevant to selling applications. Owning the application or channel becomes irrelevant to selling the content. To the extent that Java or a similar language prevails, software becomes truly open for the first time. The Microsoft desktop becomes a commodity; the Intel microprocessor becomes peripheral—the key "microprocessor" is the software code in the Java interpreter.

The computer hollows out, and you no longer are concerned with its idiosyncrasies, its operating system, its instruction set, even its resident applications. Instead, you can focus on content—on the world rather than on the desktop architecture. If you want to run a helicopter model on your screen, you don't have to worry about whether you have AutoCAD on your hard drive. You can run a video of the helicopter without owning the right decoder. The helicopter flies over the net with its own executable code. The network is no longer a threatening place. If you want to use a program from Finland, you don't have to worry that it will introduce a malignant virus to your machine.

Your computer will never be the same. No longer will the features of the desktop decide the features of the machine. No longer will the size of your hard drive or the database in your LAN server determine the reach of your information processing. No longer will the programs in your machine determine the functions you can perform. The network is the computer. The computer becomes a peripheral to the Internet and the web.

Since the release of Java and a demonstration web browser called HotJava, Gosling has experienced the kind of sudden Internet celebrity undergone by Andreessen and Bina. Unlike Andreessen, he seems somewhat baffled by it all. But in the long run, his (and Joy's) contribution may be greater. Browsers and servers may come and go, but successful new languages are extremely rare. Java, or something like it, is the key to a truly interactive Internet and a fully hollowed-out computer.

Gosling shakes his head at all the uproar. He now finds himself falling thousands of messages behind in his E-mail. Linked to the net by a mere T1 line, even Sun's Java server broke down under the overload. But Gosling himself seems to be holding up better. His speaking schedule has multiplied to a rate of three talks a week. Scores of people tell him they are using the language. Giant Japanese companies such as Sony and Sega besiege him. Finally, even Berners-Lee gave Java his imprimature as "one of the most exciting things happening on the web right now."

A movie buff, Gosling sees Java, Netscape, and similar programs changing the image of the cybersphere from an alien and menacing dystopia, as depicted famously in *Blade Runner,* into a realm resembling *Star Trek,* where computers are trusted tools. For a while it seemed that this transformation might favor the emergence of the amiable and gregarious young Marc Andreessen as the paladin of cyberspace, replacing the fiercely brilliant but widely feared Bill Gates.

But as Andreessen admitted from the outset, the more likely outcome was a massive move by Microsoft onto the net where the two firms would battle with scores of contenders. So it happened. Microsoft's Internet Explorer, soon integrated into the Microsoft operating system, would capture much of the market for Internet access software—the browser market begun by Netscape.

By 1999, many experts concluded that Gates had also thwarted the Java threat by embracing the language and using it to extend the proprietary sway of Microsoft. Many feared that Microsoft would "dominate the Internet" whatever that might mean. The government marshaled the forces of the Justice Department to forestall this chimera.

Meanwhile, Bill Joy remained in Aspen pursuing the Java grail. To Joy, the most exciting development was not any eclipse of Java, but its stunning success. By mid-1998, its two key sponsors, Sun and IBM, estimated that there were nearly a million adepts of the language worldwide. On the high end, Enterprise Java Beans had emerged with the promise of allowing truly distributed computing throughout a corporation, with applications running transparently on any available computer on the network. On the low end, most of the world's producers of smart cards, digital cellular phones, personal digital assistants such as the Palm [Pilot], and other portable devices were adopting Java as a basic platform.

Software was rapidly moving toward the Internet in accordance with the Java model, which Joy calls "Lego software." EggHead Software actually closed down its retail stores, with their limited shelf space dominated by Microsoft "suites," and opened a software emporium on the Internet, where shelf space is essentially infinite. Bloatware was giving way to software objects, plug-ins, applets, and other components that were downloaded from the net when needed.

Collectively there would soon be far more computing devices employing Java than Windows. The PC market would still be dominated by Windows, but the most common computers would be digital cellular phones and smart cards. By the end of 1998, Windows was even losing the market for personal digital assistants (PDAs) used as pocket organizers and E-email collectors. Java commanded 89.5 percent of the web page development tools market for companies with less than a hundred employees.

Then, at the end of 1998, Bill Joy made a further announcement. Linking all the millions of appliances would be a new Java-based system called Jini that would connect all the world's computing components to each other as Java software objects. Replacing specialized input-output and peripheral devices and their software "drivers" would be a single model where the device rather than the central processing unit and operating system would do the connecting. The computer would disaggregate into an array of components that could all interlink as Java objects on the net. The network would definitively become the computer.

Joy explained to *Wired:* "Think of a JavaTone like a telephone dialtone. The services it connects can be hardware—disk storage, for example—or software—a consumer service. If you're providing a service, you don't have to be a Java Virtual Machine—you can be a light switch, or a legacy mainframe. What we call a lookup service goes out and finds the function—or the users—that you want, and you get back a Java object, which you reconstitute after downloading the appropriate code."

I met Joy in August 1998 at the Matsuhisa Restaurant in Aspen. Explaining the significance of Jini and the continuing advance of Java, he removed from his pocket an array of devices: a Palm [Pilot] III, a StarTac cellular phone, and a Motorola PageWriter 2000 linked to the world by SkyTel's paging technology. After demonstrating the prowess of his index finger, writing addresses on the Palm III "Grafitti" handwriting recognizer, he turned on his other two devices. With his Motorola PageWriter, affectionately called the "Barbie computer," he sent E-mails to me and my son Richard. Then he pointed out that the various "communicators" contained three separate address lists and could not interconnect.

As the different devices booted up, each tapped a different connection: two looked for the network, one for its organizer program. "There will be thousands of appliances like these," he said. "At present they cannot interconnect. Current operating software does not help them. They do not look first for their disk as a desktop machine does. Yet that is the essential history of operating systems. They begin by tapping the disk drive. Over the years they have collected scores of drivers that connect them to printers, networks, and other outside systems. But in a world of single-chip systems, it will be impossible to include all these drivers on the microprocessor. It will be necessary to give the devices the ability to link to one another. Jini treats each device as a Java object."

So who will be the next Bill Gates? To be honest, probably no one will be quite so massive a supernova. That is an anticlimax. But the men who have instigated a new era of component software and who have written the code that gets you connected, men like Andreessen, Joy, and Gosling, are the ones in a position to do it. You can count on this: There *will* be a key software player, around whose gravitational pull all other programmers must bend their light.

Chapter 11

The Storewidth Paradigm

"And David took his staff in his hand and chose him five smooth stones."

I Samuel 17:40

For the new order, the ultrawideband Sand Hill slingshot—the capitalist conjurer of the forces causing new sleeplessness in Seattle—is John Doerr of Kleiner, Perkins, Caufield & Byers. Shunning Herb Allen's summits of schmooze, where the entrepreneurial big-time is an "audacious" investment in Coke or NBC, Doerr epitomizes the venture capitalist as industrial demiurge.

As technology investor Roger McNamee puts it: "While other venture capitalists say, 'Let's start a company,' John says, 'Let's start an industry.'" So far, beginning at Intel in the early years, he has played a key role in launching industries in electronic design automation, RISC workstations, personal computers, financial software, multimedia and wireless pen appliances (five years too soon, but let's not quibble). His latest new industry will be the biggest yet. Launched in tandem with his KPCB colleague, Sun founder Vinod Khosla, who is taking the optical side, Doerr's new theme is the delivery of broadband Internet.

The supreme force of industrial change in the coming era will be Khosla's domain of optical bandwidth. Petabits (10 to the 15th) per seconds of traffic will explode down the paths of light into. . . . what? Windows? Smithereens? A new economy in the sky with diamonds, holograms, and transcendental trash cans along streets paved with IPO gold? To catch and cache the big bang of photonics and bring it to intelligent life on earth, the explosion of bandwidth requires a complement of storewidth. This became the domain of John Doerr.

Think of storewidth as the interface between bandwidth and storage. Storewidth is the conversion of abundant bandwidth and heterogeneous petabytes into accessible information. Accessibility, in turn, depends upon the inexorable light-speed delays that afflict the Internet. Typical messages make seventeen hops between routers before reaching their destinations. Storewidth is critical to the network software and hardware paradigm that can both overthrow the dominance of office suiteware and fit the new era of mobile and ubiquitous teleputers, with your software and data lodged not on your desktop but on the net. Today there are many storewidth pioneers, from Akamai and Digital Island to Exodus and Adero, and from Mirror Image to Mango. But the first to make an impact on this critical telecosmic space was John Doerr and @Home.

From the vertiginous launch of Sun, Lotus, and Compaq in the early 1980s to his fund gushing Go at pens, Doerr's career has seen several peaks and valleys. He has even dallied with New Age mysticism, weighing a midlife retreat from the madding bustle of Silicon Valley to contemplate the Tantra or the Tao: "I sometimes think," Doerr told *The New York Times* in 1987, "I would like to become a Buddhist monk." Sure, John. At other times, facing dire threats to the venture capital environment, he has even committed politics. In 1996, he helped fend off a mischievous California state securities law that would have disabled venturers by making them subject to class-action suits when the stock of one of their companies took a tumble. In the new century, he was trying to turn Al Gore into a capitalist tool and the public school system into an educational force. Silicon Valley wishes him luck.

A lean long-distance runner, with blondish hair, a cowlick, and hornrims, Doerr at forty-six is as bashfully all-business and frenetically bit-wise as Gates, whom he resembles. But Doerr is still flouting the gravity of middle age. He spearheads the Silicon Valley energy and creativity that is finally chastening Microsoft, and Gates knows it.

Watch John in action in early 1994 at breakfast at Il Fornaio in Palo Alto head to head with Jim Clark and you might guess that something was brewing—would it be interactive television or the Netscape browser?—or listen in on a three-hour conference call amid Marriott wildernesses on the road among Doerr, Clark, and Bill Joy in the winter of 1994. Bill Gates wondered what was said. If Doerr is into Zen mysticism, what karass was he kenning in December 1994 among the tacky booths and bins, the barkers and indoor bikinis at the Western Cable Show in Anaheim? Since then, Gates has been dabbling billions in cable. What tables were tipping in January 1995 in Doerr's Woodside home, among pizzas and pastas with his wife, Anne, and Marc Andreessen, Bill Joy, Andy Bechtolsheim, and Jim Gosling of Sun and Rick Schell (Netscape's VP of engineering)? Gates thinks Doerr is the nub of a hundred Silicon Valley conspiracies and, as revealed in the transcripts of the Microsoft antitrust trial, he wants his counsel Bill Neukom to look into it.

Why, in early 1995, was Doerr lurking around the NASA Ames Research Center at Moffett Field in Mountain View, California? Was he seeking evidence of alien IPOs, or was he merely looking for some hard-core Unix Christian libertarian netbender from outer space to levitate a new industry in Palo Alto?

AT&T venturer Thomas Judge told *Forbes ASAP*'s Nancy Rutter (aka, Mrs. Jim Clark): "In Doerr's business, you have to be on the fringes to make money, and that's where he is." Follow Doerr, however, from day to day, call to call, from Sun to Quicken to America Online to Netscape to Amazon, and you will find yourself at the fringes just as they invert into the Zen center of the sphere.

On a crisp Sunday in mid-December 1996, I catch up with him at Buck's restaurant, near his home in Woodside in Silicon Valley. Wearing a dark suit from church, Doerr checks for messages on his SkyTel pager, greets fellow venturer Bill Davidow passing by, and then opens a black briefcase full of technotoys. From across the restaurant, this venture colossus looks to be a frowsy salesman perhaps a little desperate to present his wares. He removes a Mac PowerBook 5300 and Sony speakers and lays them out on the table. Amid empty latte glasses and plates of ravioli with pesto and his own half-eaten hamburger, Doerr is ready to give a demo of the new industry—a forecast for what he was already confident would be 1997's Netscape-style IPO.

It takes a minute or so to boot up the Mac, checking through the 32 megabytes of RAM (teleputers, Doerr says, will boot up instantly from flash ROM). But from there on out, it is all immediate gratification. Click to ignite a Java web page with streaming stockmarket data, a c/net talking-head newscaster, and a volcano video from Venezuela. "Wow, look at that new PowerBook go," exults a jolly woman observing from the next table.

But Doerr doesn't notice her; he is on a rush through a world of his own. Click again and you have the *Sunnyvale Sun,* efflorescent with vivid speech, sports clips, and classified personals. Newspapers will be hot on the broadband Internet. Click on the classifieds and you can presumably meet their makers in living color.

This was a glimpse of @Home, a mere demo of Doerr's broadband Internet company that erupted on the world in 1997 with the expected multibillion IPO in July. Over 56k modems or even 284k digital subscriber lines, all such dynamic fare would be agonizingly slow to access. But using cable modems and the @Home backbone and storage technologies (storewidth) to link to the Internet, your personal computer is becoming the first fully interactive teleputer, with everything instant, full motion, and always on.

In the @Home system, virtually nothing that Doerr demonstrated on the table at Buck's would originate in the computer itself. Doerr's vision implied a revolution in the computer industry that would seize the power and initiative from Microsoft and move software from desktop suites onto the network. As Doerr understood, that would require not only broadband pipes but a new model of network storage.

It all began at Intel in the early 1980s when Doerr worked at a desk down a corridor from Bruce Ravenel. Then an architect of Intel's industry standard microprocessors, Ravenel became TCI Technology Ventures' influential chief of technology. Igniting the fuse for @Home were words between Doerr and Ravenel as the two old grads from "Noyce-Moore U." wove their way through the Western Cable Show in the first week of December 1994.

"I dare not call it an epiphany," Doerr says, "but Bruce and I were at the Motorola booth where they were showing off a sleek little $300 box the size of a modem that would enable telephone calls over a cable line." Doerr was little intrigued by the telephony application (the world is already full of functioning telephones), and telephony over cable lines would still be mostly unavailable six years later. But Doerr asked: "What would it cost to add an Ethernet port to the device, so you could link a computer to the Internet through it at up to ten megabits per second?"

"When the Motorola guy guessed, 'Maybe thirty dollars,'" Doerr recalled, "our eyes got as big as saucers." And a new industry was born.

When Doerr sees a hole in the line, he hits it hard and fast. Two weeks later, four days before Christmas 1994, he was in John Malone's office in Denver presenting a plan for a Silicon Valley startup to bring broadband Internet over cable. Likely IPO market value in two years? Three billion dollars. Malone also hits hard and fast. After a three-hour meeting, the TCI chief signed off on the venture without a qualm.

At that time the established personal computer industry, led by Bill Gates, believed in middleband rather than broadband and was dabbling defensively in TV. Gates hedged by buying WebTV. A way of providing mediocre web access over television sets, WebTV recalls Samuel Johnson's comment about a dancing dog: "You are impressed not by how well it can dance but how it can dance at all."

A middleband net will not be able to maintain its current momentum of growth and power. It will not be able to usurp television and telephony. Yet the valuations of Doerr's Internet companies depend on a continued exponential ascent in traffic. Thus many people believe Doerr's broadband revolution will ultimately fail—that televisions and telephones will still dominate our culture on into the millennium.

Milo Medin was Doerr's first weapon to break the bottleneck. He has spent most of his adult life overcoming crises on the net. Medin's fast rise began in 1987 when he led the creation of NASA's Internet. In 1993, when the net was still run by the National Science Foundation, the swarthy Serb with the spruce mustache and the piping voice and broadband gush had barged into Washington to persuade the entire government to embrace the Internet protocols (TCP/IP). Against heavy resistance, he mostly succeeded. It was this universal protocol that allowed net traffic to grow exponentially up a wall of worry. Although Medin's contribution was only part of the picture, he became the talk of the net. Doerr had to sign him up.

In launching @Home, Doerr and Medin again confronted the perennial doomsday adventists who gather on mountaintops of slightly older money and disparage the future of the net, talking crisis, overload, overhype, overvaluation. Tragedy of the Commons. The experts chimed in. From Howard Anderson of the Yankee Group to Andrew Seybold and Bob Metcalfe, leading analysts prophesied a crash.

A perfect conservative "suit" for Metcalfe, Medin has been there before. The answer to traffic jams on a narrowband net is creation of a broadband net. Don't tell him it is not technically possible. Who are you kidding? This is the age of the telecosm.

Leasing capacity from the fiber-based telcos, @Home created a new multigigabit network linking to the existing NAPs at MAE East in Tyson's Corner, Virginia, at the Sprint NAP in Pennsauken, New Jersey, and at MAE West in Mountain View, California. Most important and revolutionary, though, were Medin's storewidth initiatives.

At the heart of the @Home network is ingenious hierarchical memory management and caching to conceal the mazes of slow routers, sluggish switches, and narrowband wires that still lurk treacherously around the Internet. Indeed, when Doerr finally got through to Medin and, with Will Hearst, first proposed cable modems to him, he said they wouldn't work. There would be "impedance mismatches" with the hardware and software in the rest of the network.

"This kind of blew the air out of their tires," says Medin. "But then I told them how the system could work.

"You have to think of it as a distributed computer system. In such systems, every processor cannot access memory at once. You build caches and shared-memory protocols and you mirror and replicate a lot of the data so that it's always available locally. That's what you're going to have to do on the Internet."

In other words, the Internet is a computer on a planet. Like a computer on a motherboard, it faces severe problems of memory access. Thus Internet communications depend on ingenious hierarchical memory management, analogous to a computer's registers, buffers, and latches, its three tiers of speculative caches, its bulk troves of archives, its garbage management systems to filter and weed out redundant or dated data, and its direct-memory access controllers to bypass congested nodes.

Medin explains: "If a particular web page is popular in a particular locality, you have to have that page in the hard drive or even in RAM on a local server. You have to use the multicast capabilities of cable to broadcast popular information to all addresses. Above all, you have to make the system scalable. You have to phase in bandwidth, moving fiber links and nodes and fast storage facilities deeper into neighborhoods as demand rises." All this is perfectly possible technically, Medin assured Doerr.

By the year 2000, all the concepts that Doerr and Medin were introducing to cable in 1995 had become the central agenda of the Internet. As the technologies of bandwidth abundance spread through the network backbones, the World Wide Web would ultimately bog down in latency or delay. The fatal collision with the speed of light—resolved theoretically by Einstein in relativistic physics—is now reaching the established time-space grid of the Internet.

Over the next five years of the twenty-first century, the Einsteins of networking will find themselves forced to reconstitute the entire time-space grid of communications infrastructure in a similarly elastic four dimensions, bounded on all sides by the speed of light. The new Internet will differ as radically from the incumbent as quantum theory differed from Newtonian science.

The evidence for the clash with light speed pervades the telecosm. Lacking time to fetch instructions and data from remote memories, microprocessor boards are moving to single-chip systems (cf., LSI Logic, National Semiconductor, and Xilinx). Lacking time for messages to reach remote geosynchronous posts 23,000 miles away, satellites are moving to low-earth orbits (cf., Globalstar and Teledesic).

Lacking time for the seventeen hops between routers currently prevalent on the Internet, long-distance communications links are giving way to all optical systems.

Finally, lacking time to travel through all the remote mazes of the net, the World Wide Web is being forced into the new storewidth paradigm anticipated by Medin and Doerr.

Addressing the fertile interface between bandwidth and storage, storewidth is the accessibility of stored data—the time from browser click to delivery of the first bit of a stored object.

Doerr and Medin's vision for @Home is being realized step by step in the cable network, as @Home is absorbed by AT&T. But cable remains only a small part of the total Internet. Other companies will deliver on the promise of storewidth to accelerate the entire computer on a planet.

In a world of material abundance, the only inexorable scarcity is time. In a world of bandwidth abundance, latency is the residual delay and it is largely an effect of the speed of light. No matter how capacious the transmission pipes, how large in bits per second the data stream, the first bit in the message cannot move from source to terminal any faster than lightspeed allows, plus the time waiting in queues and buffers at all the switches or other nodes along the way.

For an analogy, consider a highway. Adding lanes (bandwidth) to a highway will hasten your trip only to the point that all traffic is accommodated. After that point of bandwidth abundance, adding new lanes will impart no new acceleration. If the new lanes spur you to speed past a police trap, your trip time may even increase.

In communications, the speed limit is the light-speed wall—latency—and the only way to mitigate it is to move the data closer to you. (On the highway, people-cache apartments advertise: "If you lived here you'd be home by now.") When latency rules and enforces a tyranny of the initial bit, bandwidth gives way, as the governing principle of the network, to storewidth. Delay stems from the sum of search time and delivery time. The key to search is the directory that tells the computer where any object is; hence the primacy of Eric Schmidt's Novell as a storewidth star. The key to delivery time when bandwidth is abundant is geodesic distance, the light-speed span between any two points.

To most Internet users, the light-speed limit seems still a secondary issue and storewidth a triviality. There are many sources of delay more acute, from router conflicts to congested T-1 lines to narrowband Internet Service Providers. But the problem of storewidth emerged dramatically in Europe as long as four years ago. As an Internet addict in Stockholm, Sweden, in 1996, Sverker Lindbo was forced to confront the dilemmas of storewidth. As a result, he came up with the essential concepts behind a world-leading storewidth company now called Mirror Image.

With Stockholm one of the world's first cities to install a fiber-optic network, Lindbo commanded a state-of-the-art Internet connection, and Sweden already boasted one of the world's highest rates of Internet use. But just as today, most Internet content was secreted in servers in the U.S. five to eight thousand miles away across the Atlantic. With undersea bandwidth links costing thousands of dollars for a one-megabit connection, Swedes could not readily access the web information being popularized by Netscape's browser.

Epitomizing the Scandinavian predicament of the time is a story Doerr likes to tell about Milo Medin. It also gives some clues as to how this young net nerd from NASA became a legend in his own time: "In 1988, a Finn—call him Lars—hacks his way into Milo's computers. Ticks Milo off. He does a trace route and finds his way back to the administrator of the domain in Finland. It's an academic site. Milo already knows Lars's IP address. You can't hide from Milo. He says to the administrator, 'We have a problem. Please have a conversation with Lars.' That upset the Finns, who say, 'We are not going to do that! We respect civil liberties here! You can post a complaint if you like, but we can't tell the guy what to do.' So Milo goes into a slow boil. Says, 'I'll give you about thirty minutes to get that guy's files off our machine.'

"Nothing happens. So Milo issues an order: 'Take down Scandinavia.' The switch is pulled. Three countries go dark. They don't notice it immediately, but pretty soon E-mail messages are not getting returned. At last, three senior administrators go to Lars, so the story goes, and they say: 'We don't care if you hack into the CIA; we don't care if you bring down NSA; and we don't mind if you abscond with all the financial bits in the Federal Reserve. But don't mess with Milo at NASA.'"

"The Finns called back Milo, said the situation had been taken care of. Milo said fine and put the service back up."

In that U.S. centered environment, Lindbo set out to engineer a way to bring the net to Sweden. Lindbo and his colleagues Alexander Vik and Paul Christen tried to persuade the large U.S. sites to set up Swedish hubs. None agreed. So Lindbo developed an entire system of transparent intelligent cacheing for transferring popular material automatically. When collocated at national access points (NAPs) and other Internet hubs around the world, the patented technique, embodied in content access points (CAPs), measures hits, caches pages, and trashes obsolete materials. With compression and replication efficiencies, 75 percent of one billion pages of the web can fit in a trove of two terabytes. Think of it like a smart database of every movie ever made. As videos are rented, it keeps track of popular titles and instantly makes more copies of *Gone*

With the Wind, High Noon, et. al., so that they are ready for instant distribution from a warehouse in every region of the world.

With the net shifting toward interactive and transactional content—originating largely in the U.S. but consumed globally—Mirror Image's special Swedish solution has become a global imperative. Internet penetration is growing far faster in Europe and Asia than in the U.S. While the U.S. share of the world's Internet users dropped from 70 percent to 54 percent in 1999, the U.S. slightly increased its share of Internet hosts (computers up full time dispensing Internet data) to 80 percent of the global total. Even with no hops or other delays, the light-speed limit alone means that Internet users outside the North American continent are at least 200 milliseconds away from the vast majority of web sites.

To fetch a web object using the Internet protocol, whether a frame, image, logo, or banner, takes two round trips between the end user and the web server. With each page comprising as many as twenty-five objects, those round trip milliseconds keep adding up even for entirely static material.

Meanwhile, more and more Internet content is dynamic, consisting of E-business, streaming audio and video, large software and data files, and interactive services such as IP telephony, group games, simulations, and other transactional items. These forms of content often require a round-trip delay of no more than 150 milliseconds to remain coherent. UUNet guarantees 80 milliseconds to its elite customers. Yet most of the Internet systems in the world show average delays between 260 milliseconds and 450 milliseconds, which rapidly multiply from milliseconds to minutes with myriad objects on an obstructed net. As Peter Sevcik has shown, Internet performance by some metrics deteriorated over the last two years of the century. The Internet business plans of thousands of dotcom companies cannot succeed unless this deterioration is abruptly overcome. Hence the storewidth paradigm.

The speed of light factor imposes four constraints: 1) It prohibits the seventeen hops among routers that the average Internet packet makes before reaching its destination. These seventeen hops consume several times the delay budget for voice and video communications, for example; 2) With web pages dispersed around the globe and multiplying at a rate of a million a day, comprehensive searches pay a global light delay tax and a complexity tax. The best search engines cover no more than 20 percent of available net contents; 3) Prevailing methods of accelerating web access do not work for the increasing share of web material that is dynamic and transactional. For example, Akamai gained a market cap of $30 billion in 2000 as a storewidth star. But its some 4,000 distributed servers concentrate on static signage, banners, frames, tables, titles, and the like; 4) Much of the promise of the web lies in liberating culture from the lowest-common-denominator programming of television and films. Current systems of distributing these multigigabyte files, often in several languages, entail making as many as ten thousand copies and sending them separately around the globe. As long as this broadcast approach prevails, the web's promise as a TV killer will not be fulfilled.

Bandwidth problems are solved by hardware. If bandwidth is inadequate to handle a particular kind of flow, such as streaming video or video teleconferencing, the best solution is to replace the pipes with larger ones and replace the routers with faster ones. The ultimate hardware solution is to create an all-optical path where software is removed entirely and information flashes around the net on wings of light.

Therefore, as bandwidth increases at the center of the net, software tends to harden into glass. Driving out the millions of lines of software code in electronic switching and routing systems, all optical networks provide a crucial prerequisite of meeting the ever rising demand for bandwidth.

As bandwidth and throughput soar, however, speed of light latency becomes an ever larger portion of round-trip time. While hardware solves bandwidth problems, it can do little to reduce the time for the first bit to be found and fetched. Speed of light latency limits must be addressed in software.

While at the center of the net software hardens, at the edge hardware must soften. Dumb telephones, TVs, and storage systems, based on simple hardwired technology, must give way to personal computers, web phones, teleputers, and smart storage that must be customized for different forms of data.

The software programs relevant to the light-speed crisis deal with searches, directories, caches, and geodesy. The first Internet accelerator that addresses all the key speed of light constraints at once is Mirror Image. The company plans thirty-two content access points (CAPs) collocated in the key national access points (NAPs) and other Internet hubs around the world. Accelerating this plan, Exodus, the fast growing Web hosting hub company headed by Ellen Hancock formerly of IBM, adopted the Mirror Image system in spring 2000 and invested $675 million in the company. Mirror Image's proprietary software, partly based on Java, manages some two terabytes of Internet data that comprises some 75 percent of web traffic (much of the rest is real-time uncacheable information, such as conversations).

Here's how it works. Every time an end user clicks on a web page, it is moved through the cache to the CAP. The second version uses a fiber to cross-connect directly between the Mirror Image router at the CAP and the ISP router. Clicks are intercepted and injected into the two terabyte storage system at the CAP. Just as your computer stores the most recent web sites you've visited so you can backtrack quickly, Mirror Image stores millions of sites used by millions of users, so they don't have to trek back overseas each time they need one.

By contrast, Akamai, for example, does not readily accommodate the dynamic portions of web pages. For a relatively small number of objects in great demand, the Akamai system is hugely effective. But the web consists of a billion pages mostly in limited demand. Even a popular site under heavy demand, such as a catalog or an encyclopedia, may have many thousands of objects. Akamai can do relatively little for such a customer.

Most important, distributed around the globe, Mirror Image's thirty-two sites eliminate the speed of light delay. Video, audio, and multimedia can be readily supplied with delays well under the obligatory 100 milliseconds. Because 75 percent of the web is on each server, searches can be swift and complete compared to the hit or miss regime currently in effect at Lycos, Hotbot, and the like.

Mirror Image's customers—ISPs and content providers—are all on one system, so that as the company's business expands, it benefits from Metcalfe's law: the more interconnected users the greater the share of Internet material will be readily available and the more valuable will be the Mirror Image service.

The familiar Sun epigram, "The network is the computer," will not suffice unless the network morphs into a colossal storage system. Today the Internet contains a trove of more than a billion web pages, growing at a pace of a million new pages a day. @Home's portal consumed 49 terabytes of storage in less than two years, according to a report from Alex Brown. Doerr's other Internet star, Amazon, consumed

42 terabytes in six months. In early 2000, Mail.com filled 28 terabytes in forty-five days. Twenty-eight terabytes approximates the total traffic per *month* on the entire Internet four years ago. Total storage attached to the net already sums up to the hundreds of petabytes; exabytes (10 to the 18th) are coming soon.

Doerr's vision is clearly coming true. Software and content are no longer held chiefly in your own hard drive. They have moved to the net, where they can be accessed by an ever greater variety of network appliances and teleputers, digital cellphones and game machines, and cheap PCs used chiefly for Internet access. Also coming true is "the hollowing out of the computer" as predicted by Eric Schmidt, the Novell CEO, then at Sun.

Making all this possible so far is one of the most unexpected and least understood technological feats of the age. For the last decade, in a colossal upset, hard drive storage technology has improved at least 50 percent faster than computing power. A miniscule spot on a magnetic or optical disk is simply easier to create than a transistor doing logic or tapping random access memory capacitors. Microprocessors are essentially two-dimensional devices that have to be interconnected by labyrinthine patterns of microscopic wire. Magnetic domains that store a bit of data on a hard drive do not have to be interlinked at all, and can be inscribed on top of one another in three dimensions.

Horribly underestimating this electro-mechanical miracle in my book *Microcosm,* I confidently predicted that "cheap and dense nonvolatile" silicon memories would displace disk drives. Instead the displacement is going in the other direction with tiny disk devices substituting for flash memories even in the narrow confines of smart cards and handheld computers. IBM in June 1999 introduced a one inch diameter drive that holds 340MB at a cost of $1.47 per megabyte, about half the cost of rival silicon flash memory. In these amazing new disk drives, the read-write heads fly above the surface of the disk at a height of just seventy-five nanometers—75 billionths of a meter is smaller than any critical dimension on the surface of a microchip—while the disk rotates at a speed of ten thousand rotations per minute (about 85 miles per hour). As has been observed, adapting for scale, this feat resembles the Concorde flying at mach speed a few inches above the surface of the ocean.

Enabled by T-1 and led by IBM and Toshiba, Quantum and Seagate, disk technology will continue to double its density every year while processor technology doubles its performance every two years. Over five years, disks should improve thirty-twofold while processors improve fivefold. General purpose Pentiums no longer shape the future of information technology. The speed of light limit favors specialized distributed processors, doing their work on location, optimized for disk access, database search, and other "thin-client" applications on the net.

The rapid gains in bandwidth and storewidth—compared to the slower advances in microprocessor MIPs—have radically changed the nature of computing. From an autonomous calculating engine commanding a few thousand bits of storage, computers have become teleputers or telestores indexing, searching, sorting, and managing hundreds of petabytes of heterogeneous files and other objects across the net. On the net, the most striking successes—Doerr favorites all—have been Netscape, creator of a browser for access to remote (or nearby) storage; Yahoo, creator of a storage-based service and portal; Inktomi, supplier of a search engine used by portals; and AOL, chiefly a huge server farm and database.

The Internet is handling not merely HTML files but a huge variety of photographs, TV shows, documentaries, documents, catalog transactions, record albums, consulting services, video conferences, multimedia courtships, newspapers, virtual mall cruises, and full bore software applications, such as Sun MAJC and Star Office. It will accommodate the full marketing and sales, viewing and trialing process of a typical commercial transaction. These simulations require massive, heterogeneous, dynamic distributed storage systems that are as capacious as today's centralized mainframe systems and as flexible as the World Wide Web.

Internet service providers, application service providers, data warehouses, E-commerce hubs, portals of all kinds, photo libraries, collocation centers, storage farms (often misleadingly termed "server farms" as if the server were still central) will comprise much of all hardware deployment. All of these storage nodes will need to be accessed by a variety of outside computers and appliances.

Driving all these changes are the imperial dynamics of the World Wide Web. In launching @Home, Medin had to face the baffling problem of heterogeneous file formats generated by different operating systems. Microsoft executives told him that this was no problem; everyone should just shift to Windows 2000. But with Unix, Linux, and mainframes hugely more stable than Windows, and Macintosh Video Editors gaining market share in these new data warehouses and repositories, the all-Microsoft solution is farther away than ever. On the World Wide Web, heterogeneous file formats will give way to IP, HTML, and its more flexible metadata successor XML, with Java a likely champion in database access. The magpies' nest of ports and interconnects, contrived to deliver speed in an era of slow networks, will surrender to various forms of Ethernet, Ethernet interface cards, and Ethernet-based ten-gigabit systems running on WDM.

The entire architecture and topology of the computer system—indeed its very existence as an integrated unit—reflects a time when bandwidth inside the computer was greater than bandwidth outside it. Today, however, not only is bandwidth outside the computer generally larger than inside, but outside bandwidth is growing some ten times faster. Such divergent deltas—different rates of change—tear systems apart and wreak new paradigms.

As the network becomes faster than I/O (input-output devices such as monitors and buses), I/O is absorbed by the network. Since I/O is the defining structure of the computer, its dissolution means that the computer disaggregates, and becomes a series of peripherals attached to the network. Your disk drive, your printer, your keyboard, and finally even your processor can be anywhere on the network. Meanwhile, the heart of the information infrastructure becomes the storage repository and the increasingly object-oriented and multimedia-centric databases it contains. Embracing all will be the World Wide Web, Java, XML, IP, and Ethernet, running on the vast boulevards of wavelength-division multiplexed optical circuits. Dumb networks and stupid storage are the smart solution for the new millennium.

As Medin and Doerr were first to see, the simultaneous explosion of bandwidth and storage dictate a similarly massive growth in web caching, a solution that paradigmatically "wastes" these two crucial abundances. But it conserves the two great scarcities of the telecosm, the speed of light and the span of life, aka the customer's time.

TELECOSMTELECOSMTELECOSMTELECOSMTELECOSMTELECOSMTELECOSMTELE-
COSMTELECOSMTELECOSMTELECOSMTELECOSMTELECOSMTELECOSMTELE-
COSMTELECOSMTELECOSMTELECOSMTELECOSMTELECOSMTELECOSMTELECOS
MTELECOSMTELECOSMTELECOSMTELECOSMTELECOSMTELECOSMTELECOSMT-
ELECOSMTELECOSMTELECOSMTELECOSMTELECOSMTELECOSMTELECOSMTELE-
COSMTELECOSMTELECOSMTELECOSMTELECOSMTELECOSMTELECOSMTELECOS

PART THREE

MTELECOSMTELECOSMTELECOSMTELECOSMTELECOSMTELECOSMTELECOSMT-
ELECOSMTELECOSMTELECOSMTELECOSMTELECOSMTELECOSMTELECOSMTELE-

REVOLT AGAINST ABUNDANCE

COSMTELECOSMTELECOSMTELECOSMTELECOSMTELECOSMTELECOSM
TELECOSMTELECOSMTELECOSMTELECOSMTELECOSMTELECOSMTELE-
COSMTELECOSMTELECOSMTELECOSMTELECOSMTELECOSMTELECOSMTELE-
COSMTELECOSMTELECOSMTELECOSMTELECOSMTELECOSMTELECOSMTELECOS
MTELECOSMTELECOSMTELECOSMTELECOSMTELECOSMTELECOSMTELECOSMT-
ELECOSMTELECOSMTELECOSMTELECOSMTELECOSMTELECOSMTELECOSMTELE-
COSMTELECOSMTELECOSMTELECOSMTELECOSMTELECOSMTELECOSMTELECOS
MTELECOSMTELECOSMTELECOSMTELECOSMTELECOSMTELECOSMTELECOSMT-
ELECOSMTELECOSMTELECOSMTELECOSMTELECOSMTELECOSMTELECOSMTELE-
COSMTELECOSMTELECOSMTELECOSMTELECOSMTELECOSMTELECOSM
TELECOSMTELECOSMTELECOSMTELECOSMTELECOSMTELECOSMTELECOSMTELE-
COSMTELECOSMTELECOSMTELECOSMTELECOSMTELECOSMTELECOSMTELE-
COSMTELECOSMTELECOSMTELECOSMTELECOSMTELECOSMTELECOSMTELECOS
MTELECOSMTELECOSMTELECOSMTELECOSMTELECOSMTELECOSMTELECOSMT-

Chapter 12

Betting Against Bandwidth

"You want to bet?" barked Intel founding father Leslie Vadasz, in the summer of 1996, interrupting my high-flying prophecy of bandwidth abundance. I tried to shrug him off with an Olympian smile. We were in Aspen, after all, in the summer, relaxed and lite in Colorado's "Bel Air North," confabulating on the perplexities of the information age for Newt Gingrich's think tank, the Progress & Freedom Foundation. Let's cool it and take lunch at the nouvelle Italian Mezzaluna, was my view. But Vadasz was serious; he wanted Popperian falsifiability with a wager, which means possible public humiliation, with real money changing hands from the information poor (me) to the silicon plutocracy.

What was Vadasz doing here anyway? I wanted to know. Senior vice president at Intel, he knew too much to trade bromides on the "digital divide" for political, academic, and lobbyist types who usually gather at think tank conferences. Vadasz had escaped the Soviet army in Budapest forty years ago at about the same time as his boss Andy Grove. The twenty-year-old Vadasz had eluded soldiers facing him on a bridge by hiding behind a truck until he could find a moment to step out and then sprint across to freedom. Making his way on his own to California, first at Fairchild and then at Intel, he had helped Federico Faggin transfer from Fairchild the crucial silicon gate technology that gave Intel its edge. Then Vadasz became part of the team that built the first microprocessor. Once bashful and poor, he was now rich and tough.

I had known Vadasz for a long time, but I always had trouble either spelling or pronouncing his name. This time, under pressure, it had started out as "Verratz" and ended in a discreet gulp. "Sure I would be willing to bet if we could settle on some terms," I said. As early as 1979, when Grove was unavailable for an interview, I had greeted Vadasz at the Intel headquarters as "Dr. Baratz." He had given me a card and, after sensing that my electronic lights were dim, had explained, with an impatient sigh, how a transistor works—and drawn for my glazing eyes a white board chart of the space charge activity in a p-n junction at the heart of the device.

Now, at this Aspen conference, I was telling Vadatz—er, Vadasz—how to run his company, offering a vision of Intel microprocessors foundering under a surging tide of bandwidth abundance. "The net will be central and the CPU will be peripheral," I declared in a ringing voice. Vadasz was having none of it. Far from peripheral, Vadasz said, Intel microprocessors would become increasingly crucial in the middleband systems that would compensate for the bandwidth famine ahead. He quoted the quip of Andy Grove: "Bandwidth rises a hundred times more slowly than our ability to use it." If broadband cannot happen fast enough, the theory goes, then we will have to compensate with more powerful chips. The microcosm will forever trump the telecosm. If I persisted in my illusions to the contrary, Vadasz wanted to get me on the record with an explicit public bet.

You got it, Les. I say that by now virtually anyone in America who wants a broadband Internet connection can get one. I define broadband as T-1 speeds (1.544 megabits per second) or higher, which offers, with compression, what is called "VCR-quality" reception. Full-motion video downstream Internet capabilities (2 to 6 megabits per second for MPEG 2, like the DirecTV satellite images) will soon be accessible almost everywhere from satellites, fiber, cable, and microwave wireless.

I focus on *availability,* not penetration. This may strike Vadasz as a copout. In his charts, he stressed the length of time before PCs would even be as pervasive in homes as TVs. I really don't care. In

2000, PCs were in 60 percent of American homes and with fast modems, that's plenty. The key issue to me is not when everyone can access bandwidth abundance but when the top 20 percent of households—comprising some 50 million people with most of the intact marriages, children, PCs, productivity, income, and wealth—begin buying broadband Internet links in volume. That is happening now. The top fifth are paying for all the false starts, bugs, and glitches that plague any new technology, thus bringing it down the learning curve where everyone else can get it a few years later at a quarter the price. That's the digital divide. The rich provide the investment and the rest reap the rewards.

In 1995, the very inventor of Ethernet and a pioneer of the Internet, Robert Metcalfe himself was also placing a peculiar bet against the telecosm. He began prophesying lugubriously into every megaphone he could grasp, from *The New York Times Magazine* and PBS to *U.S. News* and *Infoworld,* that the Internet will collapse from traffic overload in 1996. He later refined his forecast to predict that disillusioned Internet users, weary of traffic jams, will retreat to private intranets, shielded from the public system and largely unavailable to it.

Et tu, Bob?

Metcalfe, who had already cast doubt on the future of his own Ethernet, was now striking a blow against the very solar plexus of my prophecies. I had founded my confidence in the Internet on the continuing power of the law of the telecosm, an edict adapted from Metcalfe's very own law of networks. In its usually quoted terms, Metcalfe's law ordains that the effectiveness of a network rises by the square of the number of terminals attached to it. But this is merely the well-known law of network externalities, capturing the exponential rise in the value of any network device, such as a telephone, with the rise in the number of other such devices reachable by it. Metcalfe's contribution was shrewdly to add to this conventional square law the declining cost of Ethernet adapters and other network gear as the net expanded. I summed up these and other learning-curve factors by incorporating the law of the microcosm into Metcalfe's law and renaming the result the law of the telecosm. In simplest terms, it ordains that the value of a network rises by the square of the collective *power* of the computers compatibly attached to it.

As the network expands, each new computer both uses it as a resource and contributes resources to it. This is the secret of the stability of the Internet. The very process of growth that releases avalanches of new traffic onto the net precipitates a cascade of new capacity at Internet service providers. They supply new servers and routers, open new routes and pathways for data across the web, and buy new terminals and edge switches to upgrade their connections to the network access points (NAPs), the Internet supernodes that in turn exert pressure on the backbone vendors to expand their own bandwidth.

Because all these routes and resources are interlinked, they are available to absorb excess traffic caused by outages, crashes, or congestion elsewhere on the net. Because all these resources are growing in cost-effectiveness at the exponential pace of the law of the microcosm, and total available bandwidth on the net is rising at the still-faster pace of the law of the telecosm, the Internet has been able to double in size annually for some thirty years and increase its traffic between three and five times faster still, without suffering any crippling crashes. The laws of the telecosm explain why the most open computer networks will prevail. Proprietary networks lose to a worldwide web.

So I wanted to answer Metcalfe's challenge. As the apparent winner of a previous argument over the Ethernet, I thought I might have an edge (after all, Fast Ethernet, my choice for a dumb networking technology, outsells ATM, his choice for a smart networking alternative, by at least twenty to one). But when he met me on a rainy day in May at his townhouse on Boston's Beacon Street, where he can look benevolently across the Charles River at the campus of his alma mater MIT, he was loaded for Internet bear. At the peak of his influence, the smiling coverboy of June's *IEEE Spectrum* as winner of an IEEE Medal of Honor, he was fresh from a summit of hoary industry seers and titans called Monticello Fellows.

At that meeting, with such fellows as Intel founder Gordon Moore, supercomputer inventor Seymour Cray, polymath pundit-inventor Jay Forrester, and Digital Equipment computer architect Gordon Bell, now consulting with Microsoft, Metcalfe says he learned such secrets as the coming extinction of nation states, and the expiration of Moore's law in ten years, "when I retire," said Moore.

Sometime thereafter, as he recalls, he wrote a column in which "I caught myself actually hoping the Internet would collapse." "I am way out on a limb here," he says over sushi and wasabi at a restaurant near his house, "I actually told a World Wide Web conference I would eat my column if the Internet didn't collapse. . . .

"What do I mean by a collapse? Well, the FCC requires telcos to report all outages that affect more than fifty thousand lines for more than an hour. I mean something much bigger than that." I suggested that with enough raw tuna and wasabi, his column would go down fine. But Metcalfe was dead serious.

The Internet will collapse and it will be good for us, and for the net. "The collapse has a purpose. The Internet is currently in the clutches of superstition, promoted by a bio-anarchic intelligentsia, which holds that the net is wonderfully chaotic and brilliantly biological, and homeopathically self-healing by processes of natural selection and osmosis. The purpose of the collapse will be to discredit this ideology. What the Internet is—surprise, surprise—is a network of computers. It needs to be managed, engineered, and financed as a network of computers rather than as an unfathomable biological organism."

Metcalfe warned that "back when Internet backbones carried 15 terabytes of traffic per month, the world's Ethernet capacity was 15 exabytes per month, a million times higher." (Exabytes add up. While a terabyte is a one with twelve zeros after it, an exabyte commands eighteen zeros.) "Just a small shift in local traffic onto the public net can create catastrophic cascades of congestion."

With private Ethernets increasingly becoming Intranets that can use the Internet but shield their resources from it by "firewalls," the likelihood of a crippling cascade from private to public nets grew more acute every day. According to Metcalfe, one way or another, such a disaster was now at hand.

"I ask my readers [at *InfoWorld*], and they tell me they think the net has already collapsed.

"One thing [the net] definitely needs," sums up Metcalfe, "is more people in suits." In Metcalfe's view, the trouble with the key professional associations that run the net was that they are full of bio-mystics with big beards and Birkenstocks who look like Bob Metcalfe did when he finally got his Ph.D. from Harvard after a dramatic setback the year before, when his thesis board flunked him at the last minute.

Republished under the title *Packet Communication,* with a new introduction from Metcalfe, his once-rejected thesis is now recognized as a classic text on networking that anticipated most of the evolution from the ARPAnet to the Internet. In the front of the new edition is a picture of Metcalfe as a newbie at his 1973 Harvard commencement, with a big beard and a weird shirt and off kilter jacket, looking kind of like a bio-anarchic, Harvard-hating, Hawaiian homeopath himself.

What does this all mean? The conversion of Bill Gates into an Internet obsessive and Java-clept. The jeremiads of Metcalfe, one of my favorite people in the industry, as both a technical seer and conservative economic voice in a webby-minded wilderness. What do I make of the only slothful emergence of several of my favored technologies of the past (such as fiber to the home and DirectPC satellite Internet access) and the admitted biodegradation of the net?

In an inaugural *Forbes ASAP* interview, management prophet Tom Peters predicted that the nineties would see a fabulous unfolding of new technology, accompanied with increasing outbreaks of technophobia and Marxism. Third Wave futurist Alvin Toffler greeted the initial readers of *Wired* with a similar prophecy of networked marvels that would be foiled by a multifront war against the new economy. Even Bill Gates himself greeted 1997 by predicting a new "backlash against the Internet" caused by "wild promises" that will not be fulfilled in the near future.

Even John Malone, whose TCI Corporation has spearheaded the crucial adoption of Internet cable modems, told *The Wall Street Journal* as recently as 1996 that the concept of a broadband cable Internet is still "experimental" and showing "diminishing potential revenues." "My job is to prick the bubble and get real," he said. Many of these leaders question the basic premise of communications abundance. Betting on continued communications scarcity, they apply their formidable minds to a vain effort to return from the telecosm once more to the wells of the microcosm. There they hope to find sources of renewal for a desktop technology they see as hurtling toward a bandwidth bottleneck.

For all their wealth and genius, for all their mastery of the central technologies of our age, they come together in a strange revolt against the telecosm. The power of the new spectronics looms too immense in petabits per second, and is moving too much faster than their own technologies, for these heroes of the computer age even to see it.

The failure of these men to come to terms with the immense changes ahead is echoed in the ultimate redoubts of reaction in the U.S. Congress, where the regional Bell operating companies (RBOCs) and cable television operators are seen as monopoly threats and local broadcasters emerge as paladins of free political debate.

U.S. Senator Ted Stevens of Alaska wants to know: With the deregulation of telecommunications of the millennium, who will bring links to Unalakleet, to Aleknagik, and to Sleetmute? Who will bring five hundred TV channels up the Yukon with the salmon to the people in Beaver? What will happen to the Yupik, the Inupiat, and the Inuit? Will we leave them stranded in the snow while the world zooms off to new riches on an information superhighway, or what ever you want to call it?

A senior Republican on the Senate Commerce Subcommittee on Communications, Stevens is a key figure in the telecom deregulation debate that continues on Capitol Hill in the wake of the debacle of a million words of partial deregulation in 1996. As he contemplates the issues, he has reason to be suspicious of the grand claims of an information age. He knows that universal service—the magic of dial tone that most Americans have taken for granted in their own homes for decades—has hardly reached rural Alaska at all.

In Beaver, Alaska, for example, there is one telephone in a hut linked to a nine-foot satellite dish. Permafrost and cold economic reality make it impossible to extend dial tone to the several hundred households of this town, even though its average household income, mostly from salmon fishing, is around $120,000. In general, copper lines in rural areas cost between ten and thirty times as much per customer as they do in cities. Without some $20 billion in cross-subsidies whereby urbanites pay for phone service in the boondocks—such as Aspen or Tyringham—rural politicians believe that their constituents would be too information poor even to vote.

Portentously sharing Stevens's concern are other powerful senators from rural states. Extended now to the Internet, their concerns could pose a deadly obstacle to true deregulation of communications and thus to continued American leadership in these central technologies of the age. At stake is some $10 trillion of potential value to the U.S. economy.

Lending huge physical authority to the Sisyphean socialist scheme of cross-subsidies and "universal service funds" is the copper wire itself, some 65,049,600 tons of it rooted deep in the rights of way, depreciation schedules, balance sheets, mental processes, and corporate cultures of the remaining regional Bell operating companies (RBOCs) and other so-called local-exchange carriers. The minimum replacement cost of these lines deployed over the last fifty years or more—and still being installed at a rate of at least five million lines a year—is some $300 billion.

In this cage of twisted copper wires writhe not only the executives of the telephone companies, but also the addled armies of telecommunications regulators, from the Federal Communications Commission and other Washington bodies to fifty state public utilities commissions and the towering hives of lawyers in the communications bar. The coils of copper also subtly penetrate the thought processes of MIT Media Lab gurus, libertarian lobbyists from the Electronic Frontier Foundation, and myriad political analysts who see this massive metal millstone as a fell weapon of monopoly power. The copper colossus even intimidates scores of staunch Republicans who arrived in Washington determined to extirpate every government excess, but who now bow before the totem of universal service socialism in their districts. You have met Senator Stevens. He has many friends on both sides of the aisle. To all these forces, local telephone service is a natural monopoly built upon a paradigm of expensive copper and scarce bandwidth.

Like any socialist system, the copper colossus will die hard. But die it must. The telco and political revolt against the telecosm will fall before the rise of spectronics: the waves of change in communications released by the magic of Maxwell's rainbow.

Some twenty years ago, AT&T's long-distance lines—like the RBOCs' local lines today—comprised an imperious cage of copper wires, installed over the previous century and similarly impossible for rivals to duplicate. Then, too, telecommunications analysts termed telephony a natural monopoly because the system could handle additional calls for essentially zero incremental cost and because network externalities ensured that more customers meant an exponentially more valuable system. These assumptions had led to government endorsement of the old nationwide Bell monopoly as a common carrier committed to universal service and subject to extensive regulation.

Regulators, politicians, and litigators always imagine that they can promote various high-minded goals, such as universal service, by offering monopoly privileges to companies for providing it. But the result, as Peter Huber and his associates show in their text, *Federal Broadband Law,* is mostly to promote monopoly at the expense of real universal service, which ultimately depends not on law but on innovation. As a form of tax, government regulations naturally reduce the supply of the taxed output. It is technological and entrepreneurial progress, impelled by deregulation and low tax rates, that brings once-rare products into the reach of the poor, who are always the world's largest untapped market.

In the case of long-distance telephony, the decline and fall of the AT&T monopoly was not chiefly an effect of politics or litigation but of technology. The millimeter waves of microwave radio effectively dissolved the long-distance copper cage. It turned out you could set up microwave towers anywhere and provide long-distance services at radically lower cost without installing any new wires at all. Finally this recognition even reached the FCC and in 1971 it authorized MCI (Microwave Communications Incorporated) to compete directly with AT&T. Within less than a decade, MCI added to its panoply of aerial microwaves the yet more advanced technology of single-mode glass fibers. AT&T's "natural" monopoly in long distance was a thing of the past.

Today, the monopolies in local phone service face a threat from spectronics technology still more devastating than the microwave and fiber threats to AT&T's long-distance empire. As with microwaves in long-distance, the government—in the name of *preventing* monopoly—dallied for decades before acting to allow elimination of the monopolies it had earlier established. After the invention of cellular at Bell Labs in 1947, some thirty-four years passed in Washington before the FCC finally began granting licenses to use it. However, when the FCC and the courts finally permitted limited competition in wireless telephony, half the metropolitan licenses went to the new regional Bells, which had no interest in using wireless to attack their local loop monopolies.

As a result, the idea persists that wireless and fiber bypass telephony are expensive supplements to the existing copper colossus rather than cheap and deadly rivals to it. The installed base of twisted-pair wire still appears to many to be a barrier to entry for new competitors in the local loop, rather than what it actually is: a barrier to the entry of the regional Bells into modern communications markets. The conventional wisdom sees the electromagnetic spectrum as a scarce resource rather than a less costly and technically superior alternative to the local loop.

To comprehend the wavy ways out of the copper cage, we must go back and listen to the technology.

At the foundation of the information economy, from computers to telephony, is the microcosm of semiconductor electronics. It reaches out in a fractal filigree of wires and switches that repeat their network patterns at every level, from the half-adder at the core of a calculator chip to the mazes of switched and routed lines in the global Internet. Computer engineers lay out the wires and switches across the tiny silicon substrates of microchips. Telecommunications engineers lay out the wires and switches across the enormous silicon substrates of continents and sea beds. But it is essentially the same technology, governed by quantum science and electrical circuit theory.

Wires may seem more solid and reliable than air, but the distinction between them is largely spurious. Wires and the atmosphere are alternative media, and to the individual electron or photon, they are only arbitrarily distinguishable. Whether insulated by air or by plastic, both offer resistance, capacitance, inductance, noise, and interference as waves travel through them. In thinking about communications, the concept of solidity is mostly a distraction.

In a global marketplace increasingly unified by light-speed telecommunications, we focus at our peril on solid states, physical resources, and bounded national economies. Conceived as a continuous span of waves and frequencies tossing and cresting, reflecting, diffusing, superposing, and interfering, the telecosm is at once the most practical resource and most profound metaphor for the global information economy.

Is the spectrum a domain of limits, to be husbanded by governments and allocated by auction at a price of billions of dollars for a tiny span of wiggle rates? Is it a finite and unrenewable resource? Or is it essentially infinite and renewable? As long as congressmen believe in scarce spectrum they will continue to disrupt the telecosm in order to favor their cherished protectorate of phone companies and TV stations.

As Peter Huber explains, "Many critics believe that phone users are the regional Bells' [RBOCs] customers. But the real customers of the RBOCs—the people they have to please in order to flourish—are not the purchasers of their services, but the U.S. Congress, the FCC, and fifty state public utilities commissions. That's just the way it is under the law, and unless you can rescind the law that situation won't change."

Whenever you see a TV broadcast tower, think of a belching industrial-era smokestack. Not only is the television content wantonly pollutant, so are the powerful electromagnetic waves that carry the trash to market in untouchable but sorely obsolete analog form. Yet TV stations play a key role in every congressional campaign. The only possible way to beat either the RBOCs or the broadcasters is through political giveaways that would allow them to use their installed base and their assigned spectrum for anything they want, including Internet services and long-distance telephony. Yet critics, such as Bob Metcalfe, prefer to storm in righteous futility against the regional Bells. I say give in. Most of the broadcasters and regional Bells lack the entrepreneurial agility to prevail in the future regardless of how many advantages they are awarded. The success of any that manage to escape from the copper cage can only advance the telecosm.

Even the FCC, which commands more technical sophistication than the rest of Washington, has adopted the prevailing paradigm of scarcity that views bandwidth as a dwindling asset to be hoarded. At a time when the world could be taking to information highways in the sky—plied by low-powered, pollution-free teleputers—the FCC is building a legal infrastructure and protectionist program for information smokestacks and gas guzzlers.

For example, since 1994, the FCC has been auctioning spectrum to the highest bidders. The auctions certainly seem preferable to the agency's previous policy of simply giving it away to "worthy" applicants among radio and television stations and established telcos. Their aim is to place spectrum into the hands of the highest bidders, who would have a financial incentive to use it productively. The procedure is slated to raise some $50 billion in revenues for the government. Yet these auctions tend to produce a winner's curse. Fifty billion dollars is close to twice the total annual investment of the telcos in new plant and equipment. As Eli Noam comments: "We can agree that the government should not subsidize telecom with an industrial policy. Can we not also agree that government should not impose crippling special fees on the industry?"

Still more subversive of good policy, the very auction process entrenches obsolescent technology and promotes the false idea that spectrum is the basis of a natural monopoly. Using real estate imagery, the FCC and many communications analysts describe spectrum as "beachfront property" and the perpetual auctions as a "land rush." They assume that radio frequencies are like analog telephone circuits: no two users can occupy the same spot of spectrum at the same time. Whether large 50-kilowatt broadcast stations booming Rush Limbaugh's voice over fifteen midwestern states or milliwatt cellular phones beaming love murmurs to a nearby base station, radio transmitters are assumed by the FCC to be infectious, high-powered, and blind. If one is driving on the frequency highway, everyone else has to clear out. Both the prevailing wisdom and the entrenched technology dictate that every transmitter be quarantined in its own spectrum slot.

Yet the beachfront analogy is absurd. Imagine, for the sake of argument, that it is 1971 and you are chairman of the new Federal Computer Commission. This commission has been established to regulate the natural monopoly of computer technology as summed up in Grosch's law. IBM engineer Herbert Grosch in 1956 had decreed, in essence, that computer power rises by the square of its cost and thus necessarily gravitates to the most costly machines.

According to a famous IBM projection, based on Grosch's law, the entire world could operate with only fifty-five mainframes, accessed through time-sharing from dumb terminals and keypunch machines. The owners of these machines would rule the world of information in an ascendant information age. By the Orwellian dawn of 1984, Big

Br'er IBM would have established a new digital tyranny, with an elite made up of the data-rich dominating the data-poor.

As head of the computer commission, you launch a bold program to forestall this grim outcome. Under a congressional mandate to promote competition for IBM and thus ensure the principle of universal computer service despite the IBM monopoly, you establish seven Bell-style licenses in each metropolitan major trading area (MTA) and seven more to serve every rural basic trading area (BTA). To guarantee universal computer service, you mandate the free distribution of key-punch machines to all businesses and households so that they can access the local mainframe centers. You plan to auction off the licenses to the highest bidder.

You never notice—because no one calls it to your attention—that a tiny company called Intel in Mountain View, California, was about to announce three new technologies together with some hype about "a new era in integrated electronics." After all, these technologies—the microprocessor, the erasable, programmable read-only memory (EPROM), and a dynamic random access memory—are far too primitive even to compare with the massive and powerful machines from IBM that seem to control the computer market in 1971.

The likely results of such a Federal Computer Commission policy are not merely matters of conjecture. The French government actually more or less did it during the late 1970s when it distributed free Minitel terminals to its citizens to provide them equal access to government mainframes. While the United States made personal computers nearly ubiquitous—with some 200 million sold over the last two decades—and launched the Internet, the French chatted away on their Minitels through central databases and ended up with one quarter as many computers per capita as in this country and one tenth as many computer networks. Entering the new millennium, the PC and the Internet are spearheading the globally dominant U.S. economy, while France founders without a single major computer company or software firm.

The FCC's spectrum auctions are exactly analogous to this hypothetical scenario. Rather than selling exclusive mainframe licences, the current FCC has been selling exclusive ten-year licenses to about 2,500 shards of the radio spectrum. Meanwhile, although no obvious

Intel of radio has emerged to overthrow this mainframe paradigm (Intel's eventual chip domination was not obvious in 1971 either), an array of companies—with names such as Qualcomm, Samsung, Nokia, Tensilica, Harris, Analog Devices, Blue Wave Systems, Tadiron, Tantivy, Quicksilver, and Morphics Technology Inc.—are introducing the microprocessors of the radio business. The software radios they are designing are potentially as revolutionary as Intel's microprocessor, showing the way to transform the entire world of wireless communications. In order to function, they do not require any assignments of spectrum.

In time, these digital radios will have an impact similar to that of the personal computer. They will drive the creation of a cornucopia of new mobile services—from plain old voice telephony to wireless video conferencing—based on ubiquitous networks in the air. Endowing Americans with universal mobile access to the World Wide Web, these devices will spearhead a new generation of computer-led growth in the U.S. economy. Eventually, the implications of smart radios and other major innovations in wireless will crash in on the legalistic scene at the FCC.

And that's only the beginning of the story.

Existing cellular systems operate in a total spectrum space of fifty megahertz in two frequency bands near the 800-megahertz level. Personal Communications Services (PCS) uses four times that space in a frequency band near two gigahertz in microwave territory. But still higher frequencies allow use of lower-power radios with smaller antennas and longer lasting batteries. For voice telephony and mobile wireless services, the existing system of exclusive spectrum allocation will prevail for the next decade. But new services will use the new radio technologies.

Beginning with the superhigh frequency spaces above 28 gigahertz microwaves and with all the excess government spectrum now being privatized, the FCC should open up unlicensed spectrum for all to use. Not only can numerous radios operate at noninterfering levels in the same frequency band, they can also see other users' signals and move to avoid them. In baseball jargon, they can hit 'em where they ain't; in a football idiom, they run for daylight. If appropriately handled, these technologies can render spectrum not scarce but abundant.

Smart radios suggest not a beach but the endless waves of the ocean itself. You can no more lease electromagnetic waves than you can lease ocean waves. This new model is actually far more appropriate than the old for Al Gore's "information superhighway." You can use the spectrum as much as you want as long as you don't collide with anyone else or pollute it with high-powered noise or other nuisances.

In general, the FCC should not be in the business of licensing spectrum. It should instead issue driver's licenses for radios. A heavy burden of proof should fall on any service providers with blind or high-powered systems, who maintain that they cannot operate without an exclusive license, who want to build on the beach and keep everyone else out of the surf. The wireless systems of the future will offer bandwidth on demand and send their packets wherever there is room.

The spectrum that Maxwell discovered was and continues to be infinite, ubiquitous, instantaneous, and cornucopian. Infinite wave action, not the movement of inert masses, is the foundation of telecosmic physics. It is ushering in an age of boundless bandwidth beyond the dreams of most prophets still confined to the copper cage or hiding behind beachfront barricades. This expanding wavescape is the most fertile frontier of the information economy. In its actions are the essential character of the coming economics of abundance and increasing returns, not scarcity and monopoly.

In contemporary networks, as Nicholas Negroponte stresses in his best-selling book, *Being Digital,* all bits are fungible. In spectronics, all spectrum is fungible. In particular, the telecosm dissolves the distinction between wireline—the foundation of the copper cage—and wireless service. A wire is just a means of spectrum reuse. Down adjacent wires, appropriately twisted or insulated, you can transmit the same frequencies without fear of interference or noise.

Using new digital radio technologies, you can beam the same frequencies through the atmosphere, insulated by air, completely mobile, and with hugely lower installation costs. From the spectronic point of view, copper wire, fiber, air, and orbits offer technically similar paths for bits and all can contribute to the coming bandwidth explosion.

The digital future is not wired or wireless. It is spectronic and spectacular. To participate in this explosive market, all telephone companies will have to escape from the copper cages of the old paradigm into the infinite reaches of the spectrum. They will have to grasp the new laws of networks in the age of the telecosm. They will have to stop rebelling against bandwidth abundance and embrace the opportunities it offers.

Chapter 13

Tilting Against Monsters

Big government and mass media feed on fear of monsters. Reporting the economy from the shores of Loch Ness, pundits peer into the shifting murk of the marketplace and see beneath every ripple of industrial change the spectral shape of some circling shark or serpent, from which only a new bureaucracy or liberal constabulary can save us.

A hundred years ago, it was the "robber barons," the creators of the great industries of oil, steel, and finance. By constantly reducing the prices of their products, John D. Rockefeller and Andrew Carnegie, along with their financier J. P. Morgan, expanded the economy to serve middle- and lower-income customers. Still more important, they laid the industrial foundation for winning two world wars and repressing both Nazism and communism. At the same time, charged with predatory pricing, collusive marketing, "dumping," and other competitive violations, those same three acquired the image of Monsters of Monopoly who could be cited to justify new sieges of government regulation.

The regulators never have trouble finding witnesses to the looming menace. In his campaigns of creative destruction, any great capitalist provokes enough panic in the establishment to fuel the beadles with testimony to bring him down. Losing competitors, whether in oil or software, are always in the vanguard of the monster hunt, which is therefore usually launched in the name of "competition" and is designed to stop it in its tracks before anyone wins.

Now, with information technology the prime source of new riches, the search is on for new monsters to justify a further governmental campaign of regulation and taxation. The first financier fully to exploit the new technologies was Michael Milken of Drexel-Burnham. He provoked the first new-era monster hunt. Pioneering the use of embryonic computer networking technology to gain exhaustive online knowledge of junk bonds, their issuers, and their buyers, he outwitted, outmaneuvered, and outearned all his rivals before the monster hunters brought him down.

Scores of books and thousands of articles have been written on Milken, but almost none focus on the quality of his investments. The famous volumes by Connie Bruck (*The Predator's Ball*), Ben Stein (*License to Steal*), and James Stewart (*Den of Thieves*) hardly mention TCI, McCaw, Murdoch, Disney, Barnes & Noble, and other successful ventures and discuss MCI chiefly as a problem of excessive debt. They actually try to blame him for the Savings and Loan Crisis, though the few banks with junk bonds would have profited massively if they had been allowed to keep them. These books are like a tome on Michael Jordan that ignores basketball to focus on child labor excesses at the factories of Nike Corporation.

Led by Michael Jensen, however, a small group of Harvard Business School scholars have scrutinized corporate behavior during the 1980s. They laboriously appraised the results of all the leveraged buyouts, junk-bond issuance, venture capital, and other tools of corporate making and remaking. Far from personifying any "wretched excess" of the 1980s, Milken emerges in Jensen's analysis as the nemesis of another form of excess—namely, the 1970s orgies of corporate waste and conglomeration by empire builders with scarcely any ownership stake in the companies they managed.

In his presidential address to the American Finance Association in 1993, Jensen declared that the central problem of the corporation is the "agency" dilemma—the divergence between the interests of the managers and the owners of large businesses. The problem afflicts all cooperative human endeavor: The interests of the individuals always deviate at the margin from the goals of the group. But, in times of rapid technological and political transformation like today, as Jensen argues, the agency gap becomes a gulf. The structure of the corporation, the training of its engineers, the skills of its executives, the costs of its processes—all become misaligned with the realities of a new technological base, a rapidly changing state of the art, and a political environment in upheaval.

In media, for example, the old establishment of AT&T, the big three TV networks, and some 1,400 over-the-air broadcast stations was breaking down into a new formation of cable and wireless schemes. Film production was dispersing into independent units, assembled just in time for ad hoc processes of production. Affecting virtually every company in the economy and threatening most existing management plans and practices, these trends created huge opportunities for wealth creation and status disruption.

A specialist in finance, information systems, and operational research at Wharton, Milken began his career at Drexel in 1970 with a computerized move to speed up the delivery of securities to its customers, thus saving the company some $500,000 in interest charges and setting a new standard in the industry. In Los Angeles, he created an advanced system for trading based on what was then a state-of-the-art Prime 550 Model 2 "supermini." Through the RS-232 9,600-baud serial ports of up to 250 Televideo terminals, the Prime computer timeshared a Fortran database containing the trading history of all Drexel customers, some 1,700 high-yield securities, and some 8,000 securities in the public market. Salespeople could now view the name, issue, and ratings of a security and compute complex yields and cash flows involving call features, sinking funds, refund schedules, puts, warrants, and prices. Meanwhile, at rival firms, many dealers were still fumbling with the levers on their $3,000 Monroe calculators.

Most of the features of Milken's system are common today, but in 1980 they were novel. This customized $2 million computer scheme, with five times that amount for programming and maintenance, gave the Drexel team a mastery of the high-yield market that seemed sinister to outside observers and competitors. Yet it was not magic or malfeasance; it was the microcosm of the new technology joined with Milken's knowledge and investment genius.

Focusing on the turmoil unleashed by the microchip in TV, films, and telephony, Milken channeled some $26 billion into MCI, McCaw, Viacom, TCI, Time Warner, Turner, Cablevision Systems, News Corp.,

Barnes & Noble, and other cable, telecom, wireless, publishing, and entertainment companies that first began weaving together the elements of the telecosm.

With an eventual $2.5 billion from Drexel, MCI built the first national single-mode fiber-optic network, thus spurring AT&T and Sprint into action to give the U.S. a global lead in the technology. With another $1.2 billion, McCaw launched the first national wireless telephone system. And with $8 billion, TCI, Viacom, Time Warner, Cablevision Systems, and Turner, followed by many other Drexel high-yield issuers, made U.S. cable television a unique national asset, with unequaled programming and broadband links. Redeemed in the process were troubled companies providing equipment and services. One of them was Corning, which supplied 62,112 miles of state-of-the-art fiber to MCI's pioneering network—this at a time when Corning had no other customers for a crucial technology developed over the previous seventeen years. These companies and other Milken beneficiaries ended up the century worth more than one trillion dollars.

By contrast, General Motors invested $121.8 billion during the 1980s in R&D and capital equipment—while the value of the company dropped to $22.9 billion. IBM invested $101 billion while its value was dropping to $64.6 billion by the end of 1990 (on the way to what now looks to have been a trough of $41 billion in early 1995). Collectively, the five hundred largest U.S. corporations wasted hundreds of billions of dollars of free cash flow.

What was needed, according to Jensen, was a total overhaul of most of these companies, their strategic redirection, and the replacement or redeployment of roughly half of their existing managers. What occurred instead was a prodigal waste of resources in defense of the obsolete structures and practices of the incumbent management.

Using venture debt and an array of complex securities, Michael Milken overcame the agency gulf by channeling billions of dollars into companies such as McCaw or MCI, largely owned by the management. For companies not owned by their executives, he funded leveraged buyouts that transformed nonowner managers feeding on free cash flow into heavy owners of equity with virtually no liquid resources. Contracted to divert all their free cash flow to the holders of high-yield debt, the new owner-managers were forced to please the capital markets in order to fund any new projects.

With the agency problem solved, these companies became global leaders. Contrary to the claims of many economists (Alan Blinder, Robert Solow, Lester Thurow, Charles Schulze, et al), productivity soared. Productivity is the amount of output per unit of input. Mired in the murky data on service-sector productivity—which was stultified by the practice of measuring most outputs by the cost of the inputs—economists still tend to miss the prodigious real growth of the 1980s and 1990s.

For example, in the brokerage and finance arenas, productivity stagnated in the statistics. Yet, during 1973 to '87, while employment only doubled in the brokerage business, the number of shares traded daily grew from 5.7 million to 63.8 million. By 1999, the three leading exchanges were trading 1.63 billion shares daily, up almost three thousandfold since 1973 with scant impact on official productivity data. While banks invested heavily in computers and ATM machines, banking productivity also was flat in the data. In a 1998 essay "Beyond the Productivity Paradox," Eric Brynjolfsson of MIT and Lorin Hitt of Wharton explain that purchases of ATMs increased the measured inputs of banks, while reducing the output, measured in the number of checks processed. ATMs saved the customer's time, a factor absent in the numbers.

Manufacturing productivity numbers misstep by assuming that products with declining prices, such as computers and communications, are also dropping in value. Also, while registering every new steel ingot, automobile, or chocolate bar, they assume that novel products represent no productivity gain at all.

These are not trivial mistakes. During the past twenty-five years, the cost of computers has dropped approximately one millionfold. Yet the Bureau of Labor Statistics shows merely an annual drop varying between 14.9 percent in 1992 and 6.7 percent in 1994. In 1998, the government finally recognized that computer prices were dipping 40 percent per year. But meanwhile, the spearhead of real price reductions shifted to a collapse in the price of communications, which is mismeasured as a modestly declining cost of long-distance voice telephony. The problem is fundamental. Most of the experts trying to explain productivity still have no reliable grasp of what is happening.

Addressing the impact of computers, Brynjolfsson and Hitt transcended the economist's "paradox" by focusing on the performance of a longitudinal database of hundreds of firms. They found that a dollar of ordinary capital was valued by the stock market at roughly a dollar, while a dollar of computer hardware was associated with some $10 of market value. As an early user of computers and as a prime investor in the telecosm, Milken was a key source of the organizational changes that have impelled economic growth over the last twenty years.

Most striking was the productivity surge in capital, as Milken, Kohlberg Kravis Roberts, Forstmann Little, and others took the vast sums trapped in old-line businesses and put them back into the markets. Not only was the productivity of the capital left behind hugely enhanced by the disciplines of restructuring, but the newly freed capital flowed into venture funds and high-yield markets where it fueled what Jensen calls "a third Industrial Revolution."

To the politicians and the prosecutors of the era, however, Milken was an exemplary monster. They proudly brought him down.

The feds next fingered Bill Gates as a putative MicroShark hiding amid the mazes of Windows and DOS. But aiming at Microsoft's integration of the Explorer browser into the operating system, the government case implied that Moore's law itself (i.e., integrating new functions on single chips) is a form of illegal bundling. By enabling ever more polymathic microprocessors, Moore's law impels the operating systems that use the instruction sets of these chips to expand their functionality.

The integration of the browser in operating system software was in principle no different from the integration in hardware microprocessors of floating point math units, graphics cards, RAM caches, and multimedia instruction set extensions. In other words, the Justice Department was attacking the basic dynamic of the microcosm. If Gates and the other monsters occasionally overstepped the bounds of business propriety, their offenses in no way justify expansion of government control and regulation into the hugely dynamic field of Internet software and hardware, now the key force in global economic growth.

Through the mid-1990s, the favorite monster target was prime Milken beneficiary, John Malone. The titan of cable and leader of Tele-Communications, Inc., he emerged in the media as an abominable snowman ranging down from TCI headquarters in the Rockies to raid and ruin rivals, terrorize politicians, and gouge his 21 million customers, 20 percent of the U.S. cable market. A billion-dollar user of

Milken's high-yield "junk" bonds, Malone became the target of choice for Loch Ness newshounds and public law pinstripes. Then-senator Albert Gore honored him as the "Darth Vader of cable." Microsoft nemesis Joel Klein of the Justice Department in 1998 called the cable industry (and by implication its leader, Malone) "one of America's worst monopolies."

In fact, for compelling reasons of business and regulations, Malone and nearly every other cable executive of the early 1990s followed a strategy of vertical integration and control: combining content and conduit in order to extract what are termed "monopoly rents" from customers. This is the model of nearly all media companies from newspapers to broadcasters. Because TCI reached more than 20 percent of all cable customers, access to the TCI conduit could heavily influence the success or failure of any cable content venture. By controlling much of the infrastructure of cable television, Malone was in a position to control much of the content and he did.

As Andy Kessler, partner at Velocity Capital in Palo Alto and *Forbes* columnist, put it, "If you want to create a cable channel, you may have to send it through Malone's bottleneck—a satellite-dish farm outside Denver. I suspect that could cost you some four million dollars in cash, or, alternatively, you can give Malone thirty percent of your company." Many content providers chose to sell to Malone. This is in essence the way Malone built up his content division Liberty Media and the content side of TCI, which together own parts of TNT (Turner Network), the Discovery Channel, American Movie Classics, Black Entertainment TV, Court TV, Encore, Starz, Family Channel, Home Shopping Network, QVC, Video Jukebox, and an array of regional sports networks. Not surprisingly, by the mid-1990s, Malone looked like an eight hundred–pound gorilla.

The alternative to the Malone model is the common carrier regime represented by the telephone companies and the Internet. In this model you build an open conduit and exercise virtually no influence on its content. Using the phone system or the Internet, people can communicate anything they want as long as they observe the technical protocols. You "dial up" any other machine connected to the network and download or upload any films, files, documents, pictures, or multimedia programs of your choosing. Although telephone companies or Internet service providers may sell or rent content if they wish, they cannot easily privilege their own programming or police the programming of others. That is why it is perverse to make phone or Internet companies liable for crime and porn perpetrated using their systems. Their content has to compete for customers freely with all the other content available on the network. In general, the common carrier model creates not content-hoarding monsters but a content cornucopia.

The Malone model's notoriety, combined with the local cable monopolies that the government mandates, explains the pervasive hostility toward the cable industry that persists to this day. The great irony is that Malone and other cable leaders were in the process of abandoning that model at the very moment when telephone executives were avidly moving to adopt it. After Malone refused to help QVC chief Barry Diller in his bid for Paramount, Bell South offered to pitch in to the tune of $2 billion and Ameritech was also reported to be preparing a bid for the film giant. Malone himself was willing to sell his content to Bell Atlantic, led by Raymond Smith who was eager to acquire the assets of Liberty Media.

In a regime of bandwidth scarcity, the idea of combining conduit and content is valid. The owner of the conduit not only can but must control access to it—which means he also shapes the content. It doesn't matter whether the conduit company is headed by a scheming monopolist or run by a management team of Ralph Nader and Tipper Gore. Bandwidth scarcity will require the managers of the network to determine and monopolize the video programming on it.

In a broadband world, however, the most open networks will flourish and proprietary networks will wither. Content providers will naturally want to put their programming on everyone's conduits, and conduit owners will want to carry everyone's content. In the world of a broadband Internet, Paramount will not want to restrict its films to AT&T/TCI's network any more than AT&T/TCI will want to exclude films from other sources.

Malone's understanding of this fact—that his own model would soon expire in an environment of bandwidth abundance—at least partly motivated both his effort to merge with Bell Atlantic and his ultimate sale to AT&T. The law of the telecosm inexorably dictates mergers not between content and conduit, but between conduit and conduit. In particular, it mandates merging the huge fiber resources of the telephone companies—which are nine times as extensive as those of the cable industry—with cable's broadband links to fifty-seven million homes.

Obstructing such mergers in the name of competition, or antitrust enforcement, or regulatory caprice, is wantonly destructive to the future of the broadband economy. Only if federal policy continues to stop the interconnection of conduits can the Malone model gain a new lease on life.

In short, the real monster to tilt against is misguided government regulation. The U.S. today commands global dominance in computer and Internet technology. But as Andrew Grove has said, "Infinite processing power will only get you so far with limited bandwidth." The next generation of computer progress depends upon the efficient use of cable bandwidth to homes and home offices.

Government officials frequently demand a "level playing field" for what they call "competition." In this competitive model, there are always so many government protected "competitors" no one can win or make any money. What the politicians call a level playing field is deadly to capitalist competition. The level playing field goal prohibits technological rivalry, which necessarily is based on innovations and advantages that confer a temporary monopoly on their creators.

Washington sees the regional Bells (RBOCs) as leviathans spouting monopoly profits and a permanent source of campaign funds, local employment, and investment resources. Yet these regulated telcos are currently confronting a dire moment of truth. In the next decade, they could face effective nationalization under continued regulation. This is the fate of local TV broadcasters, whose every step—from the amount of political coverage to the children's hours on Saturday morning—is now shaped by government. The alternative is real competition. Like U.S. West in its merger with Qwest, the Bells could enter the entrepreneurial fray in a broadband economy.

The RBOCs' so-called monopoly profits are in fact highly questionable. Most of their copper wires and circuit switches—rapidly obsolescing by any objective standard—are being written off over decades. The real value of their $300 billion worth of plant and equipment is plummeting. As TCI's sharp and salty young COO Brendan Clouston put it, under "rate of return" regulations, telephone companies are used to pretending to make money when they are really losing it.

Cable companies, by contrast, are used to pretending to lose money when in fact they are raking it in. A standing joke around the offices of John Malone's former cable empire was to ask what Malone would do if the firm were ever to report a large profit. The answer: Fire the accountant. Indeed, TCI did not report even a cosmetic profit until the first quarter of 1993. Cable firms were financed with junk bonds and other debt. These sources of funds require investors to be compensated with interest payments (a nontaxable cost of business) rather than by dividends and capital gains, which come out of profits.

Cable companies ended up eight times more leveraged, measured by debt-equity ratios, than the telephone companies. However, driven by the demands of debt, the cable firms use their capital some two-and-a-half times more efficiently. Generating $30 billion in revenues, one fourth as much as the telcos, cable firms use only one tenth the capital. Cable cash flows are highly leveraged; reduce them by a dollar and you reduce the company's value and ability to attract investment by five dollars.

Into this highly leveraged arena trudges the real eight hundred-pound gorilla, government. Where does it sit? Wherever it wants. In April of 1993, it decided to sit on the cable industry. It plumped down its burly buttocks on John Malone's desk in the form of a seven hundred-page FCC document carrying out Congress's reregulation of cable TV under the Orwellian title the Cable Television Consumer Protection and Competition Act of 1992.

A Republican-led Congress eventually rescinded some of the more extreme features of cable reregulation, but the 1996 act forbidding the integration of cable and telco networks remains mostly in effect. TCI and its colleague companies are barred from collaborating, in the same geographic territory, with any local telephone firm. Integrating cable with the glass telco backbones would have allowed the United States to realize the Clinton/Gore administration's aim for an "information superhighway," but the anti-Malone paranoia of the middle 1990s thwarted this goal. In 1998, Malone sold TCI to AT&T for $48 billion, which will allow the integration to be achieved in a different way, several years later. The monster had gotten away at last.

Since Malone got off the hook, both Washington and the industry need a new monster. The prime candidate is Bernard Ebbers, head of WorldCom, owner of the world's fourth-largest fiber-optic network. When it appeared that this six foot four bearded yokel in a cowboy hat, by buying MCI, would take over 60 percent of Internet backbone exchanges in the U.S., the Justice Department smelled a monster. As a condition of the merger, it forced MCI to sell its Internet business to the U.K.'s Cable and Wireless. (Worldcom already owned 20 percent of the Internet backbone exchanges through its purchases of UUNet and the facilities of CompuServe and AOL).

Ebbers's model differs decisively from Malone's. Ebbers had no interest in content. His interests lay strictly in conduits. His aim was to compete with the telco oligopolies, many of them owned by foreign governments, by using Internet facilities to reduce drastically the costs, first of fax and data, and later of phone traffic. But the Justice Department looked only at the bogus 60 percent U.S. Internet "control" figure and saw danger. Once again, in the name of halting an alleged monopoly, the U.S. government repressed a competitive threat to real monopolies, from the European telcos to the U.S. Baby Bells.

At first glance, Ebbers would not seem to be in the same monster league as Milken or Malone. The Milken résumé included Berkeley and the Wharton School. Malone commanded a monster résumé even before he entered business, including two MIT master's degrees (in electrical engineering and in industry management), a Ph.D. in organizational theory, and experience at Bell Laboratories. Ebbers, by contrast, grew up in Edmonton, Alberta, and attended three distinctly second-tier colleges, flunking out of two of them. Yet his story is one of the most fascinating, improbable, and inspiring in North American business. The first boot came from the University of Alberta, where the science courses in his chosen major, physical education, proved too rigorous for him. Then he failed at Calvin College in Grand Rapids, Michigan, where he was apparently predestined to be a slacker. No sixties-style blade busy with higher thoughts than academics, Ebbers tried yet again. He won a basketball scholarship to Mississippi College, a small Christian liberal arts school in Clinton. There he finally obtained his bachelor's degree in physical education and learned from his college coach, James Allen, the key lessons of his life. As Ebbers summed it up, "With hard work, dedication, a commitment to principles, and a commitment to Jesus Christ, life can be worthwhile."

His biggest asset was his parents' poverty. Because his family had never had much money, Ebbers recalls, he could always "sleep well with debt," the one obvious trait he shared with Milken and Malone.

He began buying penny ante motels in rural Mississippi with money he borrowed against the equity in the homes of two of his employees, who trusted him as a fellow Christian. Soon he commanded a chain of nine precariously financed motels.

In order to raise cash flow to help expand his motel business, he fixed on the hyperbolic business plan of a then failing long-distance telephone reseller. Though unprofitable at the time, it held the promise of a burgeoning income with better management. Following Judge Harold Greene's breakup of AT&T—which enabled competition in long distance—the phone company was reselling AT&T WATS (wide area telephone service) lines to small businesses. Buying bandwidth and selling minutes in a business with regulated prices seemed a powerful idea. Ebbers decided to buy part of the company.

Though the income streams from motels were parlous, the motels could serve as equity investments in the booming early 1980s real estate market. He could borrow against the equity in the motels to buy a further share in the richer cash flow of the phone company. With the cash flow from the telephone venture, he could expand his motel chain and benefit from the tax-free appreciation of its real estate. But this plan foundered when the telephone company deteriorated toward bankruptcy, while Ebbers mortgaged every last toilet in the motels. He took over the phone company as a matter of financial survival.

Ebbers soon found out that telephony entails three key competences: engineering, accounting, and marketing. With his Phys. Ed. degree, he was no engineer or accountant. But he could buy these skills. Marketing, however, was something he could do. In the motel business, he had learned how to sell rooms. So he brought this skill to bear on the failing telco. He quickly saw that scale was crucial in this business. The prices you could charge were set by the market and the state public utilities commissions. The costs were determined by the volume of bandwidth; a 1.544-megabits-per-second T-1 line cost twice as much per unit of communications power as a 45-megabit-per-second T-3. As you rented or purchased larger pipes, your costs per unit dropped and you made money. The secret of success in the business was expansion: buying up contiguous regional long-distance resellers and channeling them onto your own lines leased from the big carriers such as MCI and later Wiltel. Over the next decade,

Ebbers would buy and assimilate a new telephone property nearly every year.

In 1989, Ebber's company, now called LDDS, took an important further step. It merged with a publicly held company called Advantage, a chain of waffle houses with a small long-distance arbitrage business on the side. Ebbers kept the public shell and the telco business and sold off the waffles. This made LDDS a public company, which could expand more or less tax free by swapping equity. One of his first acquisitions was the telephone business of Metromedia, which brought LDDS a large and competent sales force. But negotiating for network capacity with long-distance carriers such as MCI, Ebbers became increasingly aware of a flaw in his strategy of building a reselling empire: the carriers saved all the best deals for their own customers and charged premium rates to outsiders like himself. LDDS increasingly faced a margin crunch.

Analyzing the problem, Ebbers reached the startling conclusion that he could purchase his key supplier, Wiltel, for as much as $2.5 billion in stock, and pay for it by the resultant savings in operating costs. Originally named Williams Oil & Gas, headquartered in Tulsa, Oklahoma, Wiltel had become the nation's fourth-largest fiber network by running single-mode fiber lines down pipelines abandoned with the deregulation of natural gas in the early 1980s. It also had a European subsidiary that it called WorldCom. Buying Wiltel in August 1994, Ebbers acquired the name WorldCom and became a major facilities-based telephone carrier, rich with engineering talent. To Ebbers's surprise, the operational savings gained by having his own fiber were far larger than he had anticipated. His company's annual growth rate rose from 18 percent to the low twenties. He could link his now superb marketing resources to an ever expanding array of fiber infrastructure.

By 1996, Ebbers was a king of fiber. In the next two years, he bought MFS and Brooks Fiber and began building what would become the first pan-European fiber network. WorldCom now commanded some eighty local fiber-optic rings in the business centers of many U.S. and European cities. With MFS, moreover, he had acquired its Internet subsidiary, UUNet, which is the largest and widely judged to be the best of the ISPs (internet service providers). The companies that now make up WorldCom brought total Internet service revenues of some $1.3 billion in 1997 or more than six times AT&T's WorldNet. With Internet revenues overall growing at a quarterly rate of nearly 25 percent, WorldCom's Internet income is rapidly approaching more than half their total. Although MCI gives WorldCom a powerful stake in the existing telco long-distance establishment and cash-flow, WorldCom is perfectly positioned to shift revenues from the telcos onto the Internet through its IP, fax, data, and voice offerings. WorldCom also runs the two metropolitan access exchanges (MAEs, West and East) where first-tier Internet service providers exchange traffic with their peers. Amazingly Ebbers had no broad Internet strategy until he bought MFS and inherited UUNet. But he quickly realized that "the Internet is a gorilla which is going to take over the entire industry."

Ebbers is clearly launching the world's most menacing business challenge to the world's telecom establishment. In little more than a decade, directly competing with many of the world's largest and most politically favored companies, he transformed a failing reseller telco linked to nine shaky motels deep in rural Mississippi into the spearhead of the world's most rapidly changing and cosmopolitan industry.

Like Milken and Malone before him, Ebbers has shown the magic of entrepreneurial vision and temerity. While the establishment telcos subsidized their weaknesses and created expensive obsolescence, Ebbers saw that there are three pivotal technologies in present-day telecom—fiber, Internet, and wireless—and by 1998 he had the commanding position in two of them. In 2000, he moved to merge with Sprint and acquire Sprint PCS, the world's leading CDMA-based cellular company and a pioneer of wireless Internet services.

Furthermore, Ebbers has the power, almost singlehandedly, to bring down much of the telecom establishment, with its mindless Bell company mergers of mediocrity with bloated obsolescence, its inane trials of video on demand and interactive TV farceware, its thousands of Armani lawyers and Gucci lobbyists and public affairs consultants, its alliances with long-distance carriers and regional leviathans around the world. The establishment still holds most of the pieces: $600 billion of global telecom revenues, sixty-five million tons of copper wire in U.S. local loops, $76 billion of U.S. long-distance revenues, $10 billion in leased lines to corporations, intimate relations with governments around the globe.

It is indeed an imperial array. But they are the wrong pieces, on the wrong board. The game is Internet data over fiber optics. Even without

MCI, Ebbers has mounted a nearly impregnable position in this key field. Under John Sidgemore, WorldCom's UUNet is becoming a global leader.

The U.S. establishment telcos command an apparent capital edge of over fiftyfold, with a global-installed base of plant and equipment worth roughly $263 billion, compared with Ebbers's roughly $5 billion. Altogether the international telcos command more than $1 trillion worth of capital assets or 200 times WorldCom's stake. But look at the famous "Q factor," the ratio of market value to replacement cost as conceived by Yale economist and Nobel laureate James Tobin. It can serve as an index of the entrepreneurial dynamite in a capital stock. With a market cap of some $264 billion and a similar capital installed base, the "Q" of AT&T and the Baby Bells is approximately one. With a replacement cost of some $5.5 billion and a market cap of some 33 billion, Mr. Ebbers's pre-MCI "Q" was 6.5. The telco establishment constitutes a Maginot monopoly.

The network of the existing telco regime is optimized for voice, which requires a small allotment of bandwidth for a long time (typically 64k bits per second for a few minutes). Over the past five years, the telco establishment in the U.S. and around the globe has invested tons of billions of dollars installing tons of millions of new voice lines. Thus it incarcerated its capital and personnel ever more inexorably in million-ton cages of copper wire.

The growth of Internet traffic, up some thousandfold since April 1995, makes voice an ever diminishing share of the communications business. By 2004, this trend could reduce analog voice to near 1 percent of the total. Data transmission is drastically different from voice. Coming in unpredictable bursts, it requires large bandwidth for a short time (billions of bits per second for milliseconds) and delivered in packets. On the floods of Internet data, voice will flow as an almost imperceptible trickle.

The telcos' failure to convert their networks for packet-switched data is perhaps the greatest blunder in the history of business. As former FCC Chairman Reed Hundt put it: "What we need is a data network that can carry voice, instead of what we have today, a voice network struggling to carry data."

Like AT&T's purchase of McCaw Cellular for some $20 billion in 1994 and purchase of TCI for $48 billion in 1998, WorldCom's $50 billion purchase of MCI further vindicates Michael Milken's telecom vision of the late 1970s and early 1980s. It was then that he began a $2 billion investment program designed to transform a $280 million company called Microwave Communications Inc. into a fiber colossus. MCI eventually went astray, wasting $2 billion in a Hollywood gambit with News Corp. and dissipating its early Internet lead. Looking at MCI as a going concern, British Telecom thought it was overpriced. Looking at the company as a strategic asset for its fiber and Internet resources, Ebbers could pay twice as much as BT bid . . . and WorldCom's share prices rose as a result. That is the power of an entrepreneurial idea.

Milken demonstrated the power of entrepreneurial ideas to produce mountains of wealth during the 1980s, before he was brought down by vindictive rivals in league with government. Malone was the next target because his cable holdings threatened two powerful establishments with regulated monopolies—broadcast TV and the RBOCs. Ebbers's fiber and Internet empire stands ready to release many more trillions of dollars in wealth in Internet commerce and communications and threaten monopolies around the globe. He is a hero of the dimensions of Rockefeller and Milken. Both were ultimately balked in a government monster hunt.

The monster hunters are already after Ebbers. In a stunningly bald defense of existing monopolies, antitrust regulators have already weakened WorldCom's Internet position by forcing the divestiture of MCI's Internet business. Then they moved to stop the absorption of Sprint. It will be a major tragedy if in the name of promoting competition, government regulators further step in to stop the new fiber baron and global scourge of monopolies.

Make no mistake. The monsters of the telecosm are government regulators and politicians who don't know what to make of the Malones and Ebbers of the world. With each new ramshackle version of the Telecommunications Act, Congress inadvertently wounds key U.S. players that it should be rewarding. But in the long run, not even Congress can stop the telecosm.

TELECOSMTELECOSMTELECOSMTELECOSMTELECOSMTELECOSMTELECOSMTELE-

COSMTELECOSMTELECOSMTELECOSMTELECOSMTELECOSMTELECOSMTELE-

COSMTELECOSMTELECOSMTELECOSMTELECOSMTELECOSMTELECOSMTELECOS

MTELECOSMTELECOSMTELECOSMTELECOSMTELECOSMTELECOSMTELECOSMT-

ELECOSMTELECOSMTELECOSMTELECOSMTELECOSMTELECOSMTELECOSMTELE-

COSMTELECOSMTELECOSMTELECOSMTELECOSMTELECOSMTELECOSMTELECOS

PART FOUR

MTELECOSMTELECOSMTELECOSMTELECOSMTELECOSMTELECOSMTELECOSMT-

ELECOSMTELECOSMTELECOSMTELECOSMTELECOSMTELECOSMTELECOSMTELE-

THE TRIUMPHAL TELECOSM

COSMTELECOSMTELECOSMTELECOSMTELECOSMTELECOSMTELECOSM

TELECOSMTELECOSMTELECOSMTELECOSMTELECOSMTELECOSMTELECOSMTELE-

COSMTELECOSMTELECOSMTELECOSMTELECOSMTELECOSMTELECOSMTELE-

COSMTELECOSMTELECOSMTELECOSMTELECOSMTELECOSMTELECOSMTELECOS

MTELECOSMTELECOSMTELECOSMTELECOSMTELECOSMTELECOSMTELECOSMT-

ELECOSMTELECOSMTELECOSMTELECOSMTELECOSMTELECOSMTELECOSMTELE-

COSMTELECOSMTELECOSMTELECOSMTELECOSMTELECOSMTELECOSMTELECOS

MTELECOSMTELECOSMTELECOSMTELECOSMTELECOSMTELECOSMTELECOSMT-

ELECOSMTELECOSMTELECOSMTELECOSMTELECOSMTELECOSMTELECOSMTELE-

COSMTELECOSMTELECOSMTELECOSMTELECOSMTELECOSMTELECOSM

TELECOSMTELECOSMTELECOSMTELECOSMTELECOSMTELECOSMTELECOSMTELE-

COSMTELECOSMTELECOSMTELECOSMTELECOSMTELECOSMTELECOSMTELE-

COSMTELECOSMTELECOSMTELECOSMTELECOSMTELECOSMTELECOSMTELECOS

MTELECOSMTELECOSMTELECOSMTELECOSMTELECOSMTELECOSMTELECOSMT-

Chapter 14

The Rise of a Paradigm Star

For the last ten years, I have been looking for a successful heir to the Milken vision that launched such companies as MCI, McCaw, and TCI during the late 1970s and early 1980s. These were projects of a scale and inspiration that could change the shape of an industry, or even an economy. They were companies that became the best investment vessels of their time. But my search was in vain. As I expected, junk bonds surged back from their regulatory doldrums to become once again the nation's most profitable investment—and Leon Black made a mint trading the Drexel portfolio—but no one used high-yield securities with nearly the creativity of the master.

Then in October 1998, I was invited to speak at an investors' meeting in Cancún, Mexico, for a new fiber-optics communications company called Global Crossing. I had been impressed with its able execution of an ambitious business plan, installing fiber lines from New York to London—a venture called Atlantic Crossing—in ten months. But at the time, I did not realize that its founder, Gary Winnick, had played a key role in Milken's career. I did not realize that Global Crossing had already issued $1,250,000,000 of junk bonds, gearing up the company's equity into an entrepreneurial juggernaut like the Milken ventures. Discovering who Winnick was lent new glamour to the company.

Son of a vendor of restaurant gear who went bankrupt and died in his early fifties, grandson of a Jewish immigrant with a pushcart on the Bowery, Winnick shared with Milken a sense of mortality and mission. He also shared a sense of estrangement from the patrimonial paths of wealth that had been tread by many partners at Drexel-Burnham and Company.

At the time he first encountered Milken in 1973, Winnick was a burly young man with sideburns, a mustache, and an attitude, selling bonds to small banks and insurance companies for Drexel. A business graduate of C.W. Post in Long Island, his goal in life, inscribed on a note card he kept in his wallet, was one day to make fifty thousand dollars a year. "That would make me content," he wrote. Milken taught him the possibility of discontent making close to $100,000 every day.

Winnick recalls his first impressions of Mike: "Other guys at Drexel-Burnham were always getting up and gabbing, socializing. Mike had a tight-knit group working from seven to seven, with lunch in a bag, always buried in their phones, often two at a time. There was a tension in the room. By 1973, Milken's presence permeated the company. A lot of people didn't grasp what he was doing or why he was making so much money. They resented his influence."

Selling high-grade corporate bonds at the time, Winnick one day early in 1973 received an ADP (automatic data processing company) printout of commissions. "It was fifteen times too high. I looked at it again and saw it was for one of Mike's guys." Winnick decided then and there that he was on the wrong side of the company. He began keeping Milken's seven to seven hours and left every night, like Milken, with a briefcase full of papers. Moving to Beverly Hills in 1978, Milken chose Winnick as one of his disciples.

Winnick became his right-hand man at the legendary X-shaped trading desk in Beverly Hills where Milken rewrote the rules and transcended the limits of previous financial history. Indeed, it was Winnick who suggested to Milken the idea of an X-shaped desk ("So everyone could be near Mike").

The seeds for Global Crossing were sown in 1980, in a small conference room near that trading desk, with an ashtray in the center piled with smouldering butts, where Winnick met Bill McGowan of MCI. On the surface, McGowan was not impressive. He was sweating heavily, pacing back and forth, smoking compulsively. His business plan was to compete head to head with AT&T. Winnick did not then understand microwave technology or fiber. But he looked into McGowan's eyes and he looked "very determined." He felt a radiance of vision and conviction that echoed Milken's own. With Milken's support, Winnick initiated the first loan to McGowan of $20 million. Over the next five years, Drexel would loan him another $2 billion to build the nation's first state-of-the-art fiber-optic network at a time when AT&T experts were planning to delay the technology into the next century.

"Like some others I have known," Winnick says, implying Milken, "McGowan was a guy who had never learned the words no or stop. They just didn't fit in his vocabulary." Winnick admired McGowan, who died of a massive heart attack three years ago, but he hoped he himself was different. He hoped he would know when to stop.

Six years later, in 1985, he was driving on the Maine Turnpike to visit his older boy, Adam, at camp near Augusta. In the car was his wife, Karen, an artist and writer, and their two other sons, Alex and Matt. (The Winnicks celebrated their twenty-fifth in 1997.) Winnick had just won a major deal, luring the Burroughs Corporation account away from Goldman-Sachs, and he was reliving the triumph in his mind. He was a king of the road. He was a master of the universe. He was earning some $2 million a month, not even including his investments. Then he saw a blue police car's lights flashing in the mirror. It struck him that he was driving thirty miles above the fifty-five-mile-an-hour limit. You're losing it, he thought to himself. Something was out of balance in his life.

He recalled the days of his father's bankruptcy in the household supply business. When Winnick was sixteen years old, the family had to leave a mansion in Roslyn that had been built by a founding investor in AT&T, and move into a ranch house built by Levitt of Levittown. His father never fully recovered from the humiliation and stress. He died of a heart attack three years later at age fifty-one. Now his son, a large man with a linebacker's build, was feeling the Drexel pressure cooker.

Financial pressures force most people to continue on. "Balance," like liberal politics, is chiefly a luxury of the rich. But Winnick was already rich and restless. He resolved to resign from the company on his return.

That was 1985. Despite some $400 million in funds from Drexel directed to Winnick's new firm, Pacific Assets, he launched no major businesses during the subsequent decade. There were reported conflicts with Milken over possible contracts with competing investment houses and over interviews with the press. It was Winnick who told James Stewart of Milken's wistful inquiry about how much money it would cost to buy all the properties visible between his Century City office and the sea. But Winnick dabbled in some communications companies and hired a savvy technologist from TRW named William Lee. In early 1997, Lee interrupted a Winnick vacation in Hawai to propose a $700 million investment in fiber optics. This became Global Crossing, the first new company from any of the Drexel alumni to capture the Milken magic.

When I called Winnick, though, he was seething over a *Business Week* story that treated him as a Milken "sidekick." He did not want to discuss his relationship with the high-yield king. He acknowledged readily that Global Crossing could not exist without his experience in Beverly Hills: "Its chemistry, its work ethic, its large scale, its financial structure, its ownership by management all stemmed from Drexel. I loved Drexel. I never scaled the heights, intellectually, entrepreneurially, that I did with the Drexel group."

Indeed, when I examined closely the character of Winnick's company, I discovered that its huge promise chiefly reflects the powerful lessons Winnick had learned at the X-shaped table. But there had also been clashes and conflicts in that crucible. Winnick had lost his balance. So had Milken. And with the arrival of prosecutors and media besotted with Ivan Boesky's self-serving testimony, the venture had all gone horribly awry. Winnick understandably did not want to open old wounds as he created wealth that in 1998, on the Forbes 400 list, would excel the wealth of his mentor.

Peter Drucker has said that the largest profits go to the company that provides the crucial missing piece that completes a system. There are two crucial missing elements in the global Internet: last-mile bandwidth and undersea fiber. Last-mile bandwidth is being offered by cable modems based on technology from such companies as Broadcom and by new wireless technologies, such as Qualcomm's megabit modems, to be embedded in their pdQ cellular phones in the next two years. The crucial undersea bandwidth will be supplied by Global Crossing.

Global Crossing is a classic Milken-form company, launched with $1 billion 185 million in various forms of high-yield debt. In 1999, Winnick raised another $3 billion in junk, enough to complete his global plan. Competing with some twenty telopolies in the undersea fiber business—all with access to so-called investment-grade credit—the company may seem overmatched and overvalued.

But the financial agility and technical prowess of Global Crossing trumped the prestige and clout of the international consortia. Targeting the critical missing link in the Telecosm, the undersea connections between the U.S. and the ever more global net, and focused on speed of execution, Global laid out Atlantic Crossing 1 faster than any Atlantic cable project in history and three months ahead of schedule. In 1999, it completed the upgrade, doubling capacity to 80 gigabits per second, eighteen months ahead of the original upgrade plan, and in early 2000 it added another 60 gigabits per second.

Its acquisition of Cable and Wireless's undersea construction unit made Global Crossing the world's premier subterranean cable contractor. With funding in hand for the entire planned 88,000-kilometer network and 18,000 kilometers in service by the end of 1999, the company is on target to become a global giant of underwater bandwidth.

The acquisition of Frontier in the U.S. provides a leading-edge terrestrial fiber network. Now run by former Malone right-hand man Leo Hindery, Frontier's Global Centers for Internet hosting are a perfect match for a company premised on the explosion in overseas net traffic. Linking Frontier's huge server farms, which house some three hundred of the top five hundred web sites, directly to Winnick's global network can drastically reduce the number of hops on a typical Internet connection from more than fifteen to one or two, reducing access time accordingly. For intercontinental net traffic, the Frontier connection makes Global Crossing even more attractive.

A pure telecosm play, Global Crossing artfully exploits the defining abundance of the era—the bandwidth explosion of wavelength division multiplexing over fiber-optic thread—to bypass fragmented national networks with their overloaded switches and optoelectronic convertors and their multilayered tariffs. Tying together a global fiber network among the world's largest cities, it transcends the technical muddle of hybrid networks and the regulatory briar patch of national telopolies and international consortia.

Using a centralized operations and maintenance support system, its integrated global network offers one-stop sales and service, connecting Europe, the U.S., Asia, and Latin America. Rather than leaving its customers stranded on the beach to make deals with local telcos, Global Crossing complements its undersea facilities with terrestrial ramps directly into fifty major urban areas.

With connections through Panama, Global Crossing will be able to link London to Tokyo without offloading the traffic to any outside carrier or tolltaker. Rather than requiring an ISP, for example, to contract with a phone company in London to reach across Atlantic cable and then contract with a telco in the U.S. to cross the American continent, before contracting with yet another carrier or two to connect to the party in Tokyo, Global Crossing's customers will have a one-stop shop. They will even be able to shift bandwidth from one to another span of the Global Crossing network.

The system consists of the Atlantic Crossing, a fourteen thousand kilometer ring running from New York to London, Amsterdam, Hamburg, and Frankfurt, and other "crossings" around the globe. The Midatlantic Crossing runs from New York, through Bermuda and St. Croix, to Miami and Panama. In Panama, this Caribbean facility links with the Pan-American Crossing, which loops through Mazatlán and Tijuana on the West Coast of Mexico to Grover Beach, California, and hence to Los Angeles and Seattle. The West Coast cities in the U.S. link to Pacific Crossing, a twenty-one-thousand–kilometer loop which connects with two Japanese cities and to Asia Crossing, which reaches other Asian destinations. Over all, the company plans an integrated global network of some eighty-eight thousand kilometers linked to state-of-the-art undersea technology.

This strategy addresses the most fertile opportunity in Internet infrastructure. According to an analysis based on the total capacity, terrestrial fiber bandwidth rose some two thousandfold between 1990 and 1999. Current plans suggest another two thousand times rise over the next three years (turbocharged by WDM). This means a four millionfold advance between 1990 and 2002 during a period when Internet traffic overall will have also risen several millionfold (from 1990s hundreds of gigabytes per month to hundreds of petabytes per month). These are admittedly raw estimates—give or take a hundred petabytes—but they convey the general picture.

Meanwhile, undersea capacity increased some forty-twofold between 1990 and 1999 and will rise another eighty-two times over the following three years. That's a total of 3,444 times. That means that between 1990 and 2002, terrestrial capacity will have increased by a thousand times more than undersea capacity. Between 2000 and 2002, existing plans suggest that terrestrial capacity is rising twenty-five times more than undersea capacity is.

Even if global Internet traffic proves to be growing at far less than the millionfold every ten years, the expansion will still be huge. While current growth comes chiefly through the spread of 56K modems, the next wave will feed on the power of cable modems and digital subscriber line links, most of them always on, running between five hundred and a thousand times faster. Not only will this technology increase the flow of bits; it will also sharply reduce the hassles and frustrations of the World Wide Wait, thus greatly spurring demand.

In the relatively torpid telecom world, eighty-two times growth in three years—the estimate for undersea fiber—might portend a bandwidth glut. From the editors of *Barron's* to leading analysts on Wall Street, skeptics see Global Crossing as a risky undertaking. Perhaps the Internet is chiefly a national network after all. Perhaps French executives were right in 1997 when they told me that "the Internet is an American fad." There is possibly a limit even to the desire of U.S. college students to ogle glossy pictures of Laetitia Casta. In any case, if necessary, emergency one-way Laetitia transmissions can be broadcast from France by satellites and stored in disk drives at U.S. ISPs.

Global Crossing's skeptics cite the *law of locality,* which ordains that network traffic is at least 80 percent local, 95 percent or more continental, and only 5 percent intercontinental. This has been true historically, because you E-mail people you know, and you know people amongst whom you live. However, even if this historical trend holds, the skeptics are wrong. Undersea bandwidth is currently under 1 percent of terrestrial bandwidth and is increasing only 4 percent as fast. This might make sense if the web were destined to remain chiefly an American phenomenon. But the Internet is a planetary utility in a world economy that is increasingly woven together.

Reporting from the 1998 ISPCON, the global confab of Internet service providers, even Jack Rickard of Boardwatch was startled by the rate of overseas growth. SBC International, an ISP from China, claimed 12 million customers in the Bejing area; a Korean ISP reported 1.2 million customers around Seoul. Together Brazil and Japan have nearly as many ISPs as the U.S. does. With some five thousand, the U.S. now represents well under one third of the world's seventeen thousand ISPs. In 1999, 41 percent of the world's 148 million Internet users were outside North America and readily reachable only through undersea routes. Between mid-1997 and early 2000, the North American share of worldwide Internet users dropped from over 80 percent to 54 percent.

With growth in the number of foreign Internet users rapidly outpacing U.S. growth, undersea traffic will grow several times faster than terrestrial traffic. Take my word for it. Over the next five years, the submarine portions of the Internet will prove to be an agonizing choke point. Thus Global Crossing has a truly cosmic position as the supplier of the missing element that completes the global system.

The company combines operational, managerial, and technological advantages. Plunking down a $2 million check on a table at Paolucci's on Mulberry Street in Manhattan, Gary Winnick persuaded William Carter, the CEO of AT&T Submarine Systems, to sign up with Global Crossing. At AT&T, Carter was responsible for overseeing the construction of fully 50 percent of all the world's underwater fiber infrastructure. Global Crossing now commands a worldbeating capacity for fast deployment of fiber-optic systems.

The company's twenty-month ascent from founding (March 1997) to its first billion in revenues (December 1998) is unexcelled in the history of enterprise. A key secret of Global Crossing's stunning first year is its managerial structure. To cite Harvard's Michael Jensen again, the fatal flaw of most large corporations is a divergence of interests between the owners and the managers. The owners want bottom-line profits while the managers want top-line revenue growth, usually from mergers, acquisitions, and extravagant spending of internal cash flow. The remedy is to make managers into owners and tie up the cash flow in debt service—in short, junk bonds and an X-shaped executive desk.

At the heart of the Global Crossing juggernaut, however, is the same Milken-designed motor that impelled the early success of MCI, TCI, and McCaw: the managers own the company, but outside investors command the cash flow from the high-yield issues. This means that the managers are slaves to the value of the equity and cannot rest on the cushions of internal cash flow. They must submit to the discipline of securing outside capital. Of course, raising funds is simplified by the willingness of the founders to put up some $70 million out of their own pockets, including $40 million from Winnick himself.

For the undersea facilities, AT&T submarine veteran Carter has mobilized leading edge wave-length division multiplexing (WDM) technology. This means they can upgrade their fibers by "lighting a lambda," or using more wavelength within the multiplexing, rather than laying new cable. Atlantic Crossing currently commands eight fiber strands with a capacity of 40 Gbps, upgradeable to 80 Gbps. Pacific Crossing will have a capacity of some 160 Gbps. This compares with the 5 Gbps limit on most other underseas systems. If anything, however, technological conservatism could prove an Achilles' heel for the company. If global Internet traffic continues to rise close to tenfold a year, Global Crossings' bandwidth will soon seem paltry. If other companies can deploy radically more capacious technology (a terabit per second will be possible in three years), they may be able to achieve a marketshare coup. With larger volumes, they could push prices down faster and capture the profits currently targeted by Global Crossing.

Winnick vows not to let this happen. The company's IPO prospectus declares plans to build another Atlantic Crossing within three years. But the company offers embarrassingly conservative estimates of the global growth of data traffic. The company has hired some of the best talent from the leading telcos, from Carter of AT&T to CEO Jack Scanlon, who served twenty-four years at AT&T before moving to Motorola. They may be prone to Telco nostalgia, cherishing the importance of voice and underestimating the Internet. Winnick will have to keep an eye on them.

Global Crossing faces competition. You have read of FLAG, OXYGEN, Genesis, TAT 14, LVLT, and other sea stories. Supported by the regional Bells, the International telopolies, and other leviathans, these ventures are far too complex and bureaucratic to compete with Gary Winnick and his team. Only Genesis, a collaboration between WorldCom and Cable & Wireless, delivering up to 80 gigabits per second across the Atlantic, is a significant player. FLAG's most notable achievement was to evoke a splendid cover story in *Wired* by the superb science fiction author Neal Stephenson of *Snow Crash* and *Cryptonomicon* fame and fortune. But its top capacity is 5 gigabits per second, which seemed big at the time. Otherwise, these projects are mostly run by elephant companies copulating nervously with one another. OXYGEN is a suitably ambitious venture, headed by Neil Tagare, but it is having trouble raising funds and is running at least two years behind Global Crossing.

Measured by its opportunity, Global Crossing long remained undervalued by the market. If the Internet is ultimately going to be a global system, with traffic distributed around the world, the network suppliers with a global reach will tend to prevail. They enjoy what are termed "line economies": the ability to reach new destinations merely by extending existing lines, rather than starting from scratch. To reach any particular location it will always be cheaper for the companies that already have global bandwidth.

Global Crossing has laid five thousand kilometers of fiber to Europe. This Atlantic Crossing system was already profitable by the third quarter of 1998, when the company reported $15 million earnings on revenues of $117 million for it. Within its first year it recovered 75 percent of the construction costs of Atlantic Crossing by selling 21 percent of the capacity to such companies as Level 3, WorldPort, and Deutche-Telecom. Global Crossing is poised to become the most capacious independent network in the world.

Just as MCI pioneered single-mode fiber in the U.S., TCI transformed cable, and McCaw launched a national wireless system, Global Crossing will pioneer the first integrated global fiber-optic network, fulfilling my decade-old prediction of a "worldwide web of glass and light."

Global Crossing's shareholders, alas, may never profit from this visionary venture. The "Milken motor" worked brilliantly in the 1980s and early '90s, financing the most important information infrastructure projects of the era with a pile of junk. Under the pressure of a global dollar deflation, however, over the past five years the cost of dollars has risen some 30 to 40 percent measured against gold, other commodities, and other currencies. With their debt now denominated in more expensive dollars, the most entrepreneurial enterprises of the day, especially huge infrastructure projects like Global Crossing, Globalstar, and Exodus, are passing into the hands of their creditors. But their creations remain as foundations for the telecosmic economy.

Chapter 15

Deluge of Dumb Bandwidth

If Gary Winnick is the leader of a group of bandwidth pioneers under the oceans, who is doing it on dry land? Who will lead the charge against the telcos, and their "smart" networks? Who will help collapse the seven-layer architecture of increasingly cumbersome, routed-and-switched systems?

At InterOp '97, at a session on backbone bandwidth, a roly-poly, swarthy Egyptian-American named Nayel Shafei, then chief technical officer of Qwest, a new Denver-based outfit, shouted like a revolutionary. Qwest was using the luminous new technology of wavelength-division multiplexing to vastly expand the capacity of fiber networks. Vibrating with energy, slightly balding, Shafei on a roll tends to go into oscillation, spreading positive feedback loops and auras, colored infrared. By the time he sat down, a few minutes later, he had turned the known world of communications on its head. In *his* world, he declared, the problem of bottlenecked network electronics became an opportunity. Bandwidth abundance, he announced, was at hand.

In staccato English suffused with rhythms and inflections from a Cairo boyhood, Shafei ran his audience through the astonishing arithmetic of Qwest: "We have a revolution going here. We are delivering more bandwidth than AT&T, WorldCom, Sprint, and MCI *put together.* These companies are all haggling about when they will offer OC-12

[622 megabits a second]; well we have OC-192 [10 gigabits a second] *available today.* We can put *sixteen* OC-192 bitstreams on every fiber; that's 160 gigabits per second. And we have two conduits in the ground that can each hold at least 96 fibers. That's *30 terabits,* over a 17,000-mile network. And it's just a start. . . . We are not interested in antiquities and natural history around here. We are exploring OC-768 and above. We are cooperating with Cisco and Juniper and Avici and other router companies to develop terabit routers—routers that can switch at OC-192 wirespeeds and higher. . . . People talk of launching Internet 2 in the future, with the help of the government. We are launching Internet 2 *today.*"

I first heard this speech not in the convention center itself, but on a conference tape a few weeks later. An InterOp veteran, I had had my own agenda in Atlanta, as the speaker at an event sponsored by Digex, a national ISP. I had also brought along my then thirteen-year-old son, Richard, and spent a lot of my time trying to spoof him past the guards, who wanted to ban him as a minor. Along with preparing and delivering the speech, I worked out at the sumptuous Peachtree Center Athletic Club, watching lithe damsels bounce around the running track while I sweltered away on the Stairstepper. I also had to explain the conference in detail to the puzzled media—CNN's "Digital Jam," and reporters from *Computer World,* the *Atlanta Constitution,* and the networking trade press. I thus had no clue about what was actually happening at the show until I returned back home to western Massachusetts and started playing the tapes, as I drove back and forth between my house and office on icy winter roads. When I first heard Shafei's speech, I almost drove onto the Stockbridge golf course reaching to rewind the tape, so I could grasp all the burst-mode rhetoric.

Shafei commands one of the most original minds I have met in the world of telecom. The cosmopolitan scion of the president of Egypt's Misr bank, he spent his youth in Cairo, Sudan, Italy, and Britain. His early education took place at the hands of Franciscan nuns, in an exotic Sudanese habitat. He attended high school in Cairo, where as a freshman in 1976 he acquired his first Casio programmable calculator and was struck by its power and efficiency. Classes were otherwise unchallenging, until a sluggish summer night when he had nothing to do and ventured into the Japanese embassy. He signed up for a course in Japanese, studied for six years, and acquired a sense of new intellectual frontiers and dimensions. It was one of five foreign languages he eventually mastered.

Under pressure from his parents, Shafei entered the University of Cairo as an engineering student and discovered the mind-expanding powers of physics. There he also met his future wife, Mona, now a pediatrician and the mother of their two children. From Cairo, he went to MIT, where he further stretched his mind with Marvin Minsky in the Artificial Intelligence Lab and in Nicholas Negroponte's Media Lab. He won his Ph.D. with a thesis on network architectures and designed a language for network management that he called Enkido, after one of the warriors in the *The Epic of Gilgamesh,* a favorite book. He ended up at Allied Signal in Virginia, where he helped design the "Clipper" encryption chip, intended to allow the FBI to read Internet messages and later mothballed. But in 1998, he had found his true home at Qwest, which was then quietly springing from the fertile vision—and billion-dollar deep pockets—of secretive one-time Southern Pacific Railroad baron Philip Anschutz, whose previous foray into telecom was a little startup called Sprint (SP, get it?).

Shafei appeared also on two other tapes, pushing his gospel of bandwidth abundance. I resolved on the spot to look up this alien terahertz terrorist, who so eloquently demolished the idea of network scarcity.

Qwest was the first of the big bandwidth companies both to pursue the lowercase ethernet paradigm (i.e., collapse the seven-layer network model) and supply the bandwidth to make it work. In its essence, Qwest began at level one, the physical layer, allowing others to supply complexities above. Qwest's WDM systems and erbium-doped amplifiers don't care what kind of packets are traveling down the pipe. Its business is chiefly gigabits not smarts, IP not IQ. But by offering unheard-of speeds, Qwest is encouraging dumber and faster traffic cops above.

In 1998, I finally fufilled my resolve to interview him. I had heard that the Qwest executive was reluctant to see me. He had read a report I had written about him and had gotten the mistaken impression that I thought he was as dumb as his network. This reaction to my ironic celebration of his brilliance raised the terrible suspicion that perhaps he could not fully grasp either *double entendre* or the dumb network paradigm. Perhaps he was not so bright after all.

By the time I called him for an interview, though, he was well beyond any resentment. He invited me to visit him at his office in Denver. There I learned that Qwest is more than a company. It is a new diluvian saga. Shafei's inspiration is *The Epic of Gilgamesh,* part of a primordial literature of transformation, a tale of the time when the gods laid down the laws of light and life. At the heart of Qwest's new silver tower on Denver's downtown Seventeenth Street is a meeting room named Enkido, like his network language, invoking the blessing of his ancient heroes on the terrestrial ventures of his firm.

Reading the epic, which parallels the biblical story of Noah, I came to understand better Shafei's visionary passion. With a warrior named Enkidu, Gilgamesh experiences a saga of transfiguration, in which the dimensions of ancient reality emerge from the flood entirely transformed. Shafei believes that we live in an era of transfiguration no less far reaching than the epoch of the flood, that telecommunications is wreaking a diluvian deliverance from time and distance, that he and his rivals in bandwidth abundance are forging a new world of radically different constraints and opportunities.

Shafei sees the layers on layers of network complexity—the proliferating litter of acronyms piling up in teetering paper towers, the tons on tons of ever expanding copper wires, the ramifying millions of lines of intelligent software switching code—as a moral cancer that must be swept away like the corrupt Babel of ancient times. The tales of Noah, Gilgamesh, Ziusudra, and a hundred other ancient heroes depict the onset of rains that forever transform human life and end with the promise of a rainbow.

Qwest's diluvian saga began with a trickle of rebellion at the very heart of the smart network paradigm, AT&T. The rebels were Joe Nacchio, head of AT&T's consumer long-distance division, and David Isenberg, a lead researcher at Bell Labs. Both left AT&T—Isenberg to become an inspirational leader in the movement to dumb networks and Nacchio to build one. When Denver oil and rail baron Philip Anschutz set up Qwest in that same year, he hired Nacchio. Anschutz had the entrepreneurial vision to sell off his rail company while retaining its real value—its rights of way—which would become the conduits for Qwest's fiber lines.

Born in Brooklyn and raised in Staten Island, son of a bartender at the Governor Clinton Hotel, Nacchio grew up on the other side of the world from Shafei in every way. To Nacchio, remote horizons were not Europe and America but the skyline of Manhattan and the streets of the other boroughs of New York. Nonetheless, like Shafei, Nacchio was always a free spirit.

At NYU, he faced "near electrocution" as he now jokes, in his pursuit of an electrical engineering degree. He set his heart on a marketing career at Proctor & Gamble after his graduation from NYU in 1970. Running late to a meeting with the P&G recruiters, he walked into the wrong room, where a friendly AT&T man awaited. By the time he got to the P&G interview, he was twelve minutes late and too flustered to make a good impression. The job offer, together with a promise to pay his college loans, came from AT&T. Beginning as "a deviant with a beard and an attitude from the 1960s," he found himself dispatched to remote Cincinnati. Off came the beard and on came a wife. He obtained a transfer to AT&T's New Jersey headquarters and remained there much of the time for two and a half contentious but mostly triumphal decades.

Nacchio understands AT&T's vulnerabilities as well as anyone in America. Technically AT&T is sitting on millions of miles of "old glass." Mostly installed in the late 1980s and optimized for 64-kilobit voice, it cannot carry the dense WDM signals that Qwest is deploying. Further, AT&T has less desirable rights of way. For huge fiber bandwidth, you want protected conduits. Much of AT&T's fiber runs across rights of way defined for microwave and coax, through farms and suburbs and around the edges of cities where telephone lines were traditionally placed. Perhaps half of AT&T's glass was merely laid in trenches without hard protection.

AT&T has an intelligent network, full of expensive Bell switches, using scores of millions of lines of computer code. Its operation support systems are optimized for voice, like everything else at AT&T that is not optimized for law and lobbying. "At the end of the day," Nacchio recalls, "Bob Allen [AT&T's former CEO] loved to hear lawyers refine nuances of antitrust." What's worse, Nacchio says, "John Zeglis," the ascendant AT&T vice president, "still thinks they have the most advanced and intelligent telephone system in the world." By contrast, as Nacchio notes, the Qwest network is mostly dumb and beautiful. So although he had risen to lead a division that contributes 95 percent of AT&T's profits and was considered a serious contender for CEO after Allen, he chose to defect to Qwest.

Qwest pushed capacity to the utmost as a strategic weapon that both confounds and deters competition and opens up the huge elasticities of lower-priced communications. It could begin its long-distance phone service at 7.5 cents a minute, at the time 25 percent below the telecos bargain basement rate. Yet even at this price, Qwest's dumb network could make as large a margin as AT&T made at its once prevalent price of 14 cents, reserved as a reward for its most loyal customers (the more inert half who do not shop for a better deal).

If you want to know on which side of the fence any communications company resides, ask its CEO whether he believes demand for telephony is elastic or inelastic—that is, whether lower prices yield higher or lower total profits. Based on exquisitely calculated marketing models, conventional wisdom among the over-the-hill telco gang is that lower prices are a perilous tactic. By contrast, the new dumb network companies are all united in their faith in positive elasticities, in their belief that lower prices can expand markets and unleash innovations that transform the business. Explains Tom Evslin, another AT&T renegade, now CEO of Internet telephony company ITXC, lower prices not only expand the existing market, they summon new markets through innovation: "You get secular gains in traffic that are multiplied by breakthroughs in new services." His company bypasses direct customer service in order to supply the fabric between the thousands of Internet service providers and others who are offering IP phone connections as clear as the telcos.

Most of these IP phone services obviate the need for even the most brilliant switches. For example, consumers access Qwest's IP long-distance service by making a local call on a normal telephone, dialing into a circuit-to-IP platform made by Newbridge Networks. In the Qwest system, the Newbridge box simply packetizes the raw, 64 kbit/s signal, and sends it via IP almost directly on the physical layer, not even bothering to compress the voice bits at all.

As MIT's Internet guru David Clark explains, Internet Protocol is now reaching its potential as "the great spanning layer" between the glass fiber below and the bit stream of ideas above. Designed from the outset as an Internet-working technology to link disparate hardware and software, dumb Internet protocol is the great enabler of intelligence at the endpoints of networks. From seven layers to just a few.

"Anybody who's got an idea, but has been frustrated by limited bandwidth, should contact me immediately," Shafei said in Qwest's early heyday. Yes! Abundant bits-in, bits-out bandwidth—to liberate innovation—this is the future of telephony. As Shafei once told Isenberg, "It's your idea—the stupid network. We're doing it!" A dumb network works roughly like a river. The water in a river, or the data in a dumb net, gets to where it must go, adaptively, with no intelligence, and no features, using self-organizing engineering principles, at virtually no cost.

Intelligence in the network actively impedes innovation. Everything affects everything. For example, until recently, your local telco could not give you Caller ID for an incoming call while you were on the phone. To fix this, Bellcore had to invent a low-tech, low-functionality, high-complexity protocol called analog display services interface. In an Internet telephony system, call waiting with Caller ID would be a no-brainer.

Similarly, for a decade, the telcos struggled to find an intelligent switch-based system to deliver multimedia video on demand. But Real Networks is solving this problem on the edge of a dumb network. Real began with "broadcasts" of audio for an audience of one. Using the RealAudio player software, you can access concerts, book readings, record libraries, radio stations, and audio on anybody's web site around the world. Moving on to "streaming video," the RealVideo player is herky-jerky today. But new bandwidth can smooth it all out. Considering how far the RealAudio player came in its first three years, we can expect great things from RealVideo. Enabling a user-defined channel for everyone, such systems will be the last nail in the coffin of "five hundred channels."

Telcos have a word for the problem of complexity created by smart networks. They call it "feature interaction." During the period of the Internet challenge, entire bulging issues of telco and IEEE technical journals were devoted to the problem of "feature interaction." Telcos dread new features because every feature must be tested with every other feature. In the intelligent network, each new feature needs a business plan, a marketing plan, a provisioning plan, an operations plan, and a maintenance plan, and these feature interaction plans interact with one another as well. Jim Nacchio and Nayel Shafei are sweeping them all away.

A human operator used to be able to set up one hundred calls an hour. Modern computer-controlled circuit switches, such as Lucent's 4ESS, can complete about 1 million calls an hour. A packet switch on a dumb network can accommodate more than 3 trillion calls per hour. Meanwhile, an all-optical lambda cross-connect from Xros or Agilent can accommodate as many calls or petabits as the network can send.

In Qwest's wake many other formidable companies are pursuing the WDM rainbow. James Crowe of Level 3, a company formed by the Kiewit construction company after its sale of MFS to WorldCom for $14 billion, announced a new $3 billion network plan covering some twenty thousand miles. Williams Telecommunications, once Williams Oil and Gas, became a telecosmic pioneer by running fiber through its natural gas pipelines and selling this network to WorldCom for $8 billion in 1996. Williams now is deploying another thirty-two thousand miles of fiber down gas pipelines. IXC, a bypass carrier now owned by Broadwing, commands a national broadband fiber network comparable in size to Qwest's. Broadwing is using Corvis systems to send 160 wavelengths down 3,000 kilometer paths without opto-electronic regeneration. GTE was purchasing 25 percent of Qwest's capacity and extending a new national fiber scheme to link BBN's "Terapops" (points of presence with a capacity of terabits per second).

Even AT&T, under Michael Armstrong, is deploying some $7 billion worth of new fiber technology together with the hybrid fiber coax cable TV plant of TCI. Following a flood at a Sprint facility in Missouri in 1993, Sprint had begun deploying WDM systems as a stopgap. Liking what they saw, the company began moving the technology through its twenty-three thousand-mile fiber network.

Nonetheless, it was Qwest, under Nacchio, that changed the rules of the game. After assembling rights of way all across the U.S. and into Mexico, it moved with dazzling speed to deploy and light at least one lambda in some eighteen thousand miles of fiber. In Europe, the Qwest/KPN partnership was on track to deploy nine thousand miles of fiber by 2001. Nacchio has also succeeded in driving revenues above $3 billion a year and garnering a market cap eight times revenues.

As time passed, Shafei began issuing signals that all was not well in his bandwidth crusade. Nacchio was succumbing to the temptations of a telco, buying long-distance carrier LCI and Bell Goliath U.S. West, shirking his duties as a paladin in the passion play of diluvian dumb networks. More crucial were technology compromises Shafei believed Qwest was making in order to reap voice revenues today that might cripple the drive to be the data network of tomorrow.

Shafei listens to the technology. To him it seemed that Qwest was no longer riding the light. The Qwest buildout used advanced Lucent TrueWave fiber and Nortel industry standard SONET switches. Advertised as WDM (wavelength-division multiplexing) systems, they were more accurately described as switches adaptable for WDM, but now used mostly with only one wavelength.

SONET was created for first-generation optical networks, which used fiber as merely a more capacious version of copper, and were premised on the relatively modest cost savings and elasticities that would result from upgrading the existing voice network. In that context, SONET is an efficient and economical system. But when the target is not a 50 percent drop in the cost per bit of carrying voice, but a several millionfold reduction in cost and increase in capacity for data, SONET becomes an extravagant, insuperable barrier.

They say that when a smaller company buys a larger one, it becomes the larger one. Nacchio once remarked that "long distance is the second most profitable business after smuggling cocaine." With LCI and U.S. West (which aimed to enter long distance), Qwest was risking addiction.

Shafei had seen enough. To his way of thinking, Qwest, Level 3, and other fiber-laying companies "managed to commit in a few months all the technical sins that it took AT&T a century to commit."

In part, he acknowledged, their behavior was understandable. Buying from the big telco suppliers such as Nortel and Lucent, the new fiber companies at the turn of the century still had no risk-free way to illuminate their networks. To break away with new technologies, they would have to place bets on new companies and innovations. Instead, they took the most conservative course, reproducing and extending the old network architectures rather than adopting new ones. But this choice forced them to forgo the most important advantages of the new optics of multiple wavelengths.

Preserving the old SONET architectures turned out to be prohibitively expensive. To light up a ten thousand- to thirteen thousand-mile network with a full complement of WDM wavelengths would cost $250 million for the first SONET lambda of 10 gigabits per second and perhaps $80 million each for the next seven. Lambda number nine would cost another $250 million. To light just one fiber thread with the 160 lambdas that Nortel offered at the end of 1999 would cost some $16 billion. And that would not even include the costs of real estate, cross-connects, add-drop muxes, and other gear needed actually to deliver service to customers along the network. WDM promised the transformations of the flood, but SONET would strangle it at the source.

SONET is so costly because it is an electronic routing system for an optical network. To determine the next move for an optical bit stream, SONET transforms the optical signal into an electronic one, reads every header on every packet, makes the appropriate switching decision, and then retransforms the signal back to optical form.

In other words, this technology is good for point-to-point backbone links and giant corporate, government, and university clients in big cities, and is too costly for everyone else. Publisher Richard Vigilante of the *Gilder Technology Report* calls it "an autistic network": "There's a lot going on in there, but noone can get it out."

Early in 2000 Nacchio, implicitly conceding some of the criticisms, announced plans to rebuild his network. For Shafei it was too late. By March 1999 he had already had enough. If Qwest was not ready for the fibersphere, not ready to surmount the kilofold wall of Internet demand, then he would have to do it himself.

In early April, he told *Forbes* writer Toni Mack, who bestrides the world of telecom from her wheelchair in Houston, that he was starting a new company. It would be called . . . what else? . . . Enkido (spelled with an "o" to avoid a conflict). As partners, he enlisted Raj Reddy, the venerable computer architect from Carnegie Mellon, and Robert Kahn, the coinventor of TCP/IP. They leased six dark fibers over some twenty thousand miles from other telcos. As initial customers, they signed up DARPA and NASA for broadband service and NBC and Fox TV for transport of uncompressed HDTV signals at 1.9 gigabits per second.

With the instinct of a disrupter, Shafei saw that he could not fulfill the promise of WDM without breaking free of existing telecom practices, suppliers, and network topologies. He knew that Lucent and Nortel would not give him the tools for a truly revolutionary attack on their most lucrative customers in the telecom establishment. For Enkido, Shafei would have to turn to outside forces.

A new network would have to provide a capacious broadband backbone. It would have to create efficient access points, new central offices, in major cities. It would have to solve the last-mile problem of access to the network from the vast majority of businesses and homes that were not linked by cable modems or by the early deployments of digital subscriber lines. It would have to master a billing system to charge for its services. It would have to accommodate secure and robust Internet commerce. And, Shafei resolved, it would have to accomplish all these goals at a cost perhaps tenfold cheaper than the large telco solutions.

Starting with the backbone, Shafei went for the existing abundance, dark fiber—all the fiber that companies had wantonly laid but not lit. "There are many companies out there who have buried fiber and now are trying to figure out what to do with it," he said. Beyond Qwest and the other fiber barons, the companies with the most dark fiber were the power companies that deployed fiber for telemetry and were entering the communications business after deregulation. Combining both categories was Williams Communications, and the Tulsa company was willing to lease Enkido six fibers over several thousands of miles of their network, mostly along natural gas pipelines. Among other companies leasing fiber to Enkido were IXC, Enron, Montana Power, and a New England optical network company called Neon.

From such sources, Shafei managed to cobble together a six-fiber network covering some twenty thousand miles for a projected total of around $200 million. Strictly speaking this was not a network, since it had less communications power than two tin cans and a ball of string. And Shafei was willing to buy it at some six times as much money as it cost to build, which was a bad start.

Shafei was able to bid around six times the cost of the network because all the rest of his equipment—which represented some 98 percent of total network expense—would cost him between ten and twenty times less than the bundled "turnkey" solutions offered by Nortel and Lucent. His optics chief, Ron Haigh from Qwest, had built a cheap broadband network for the military several years ago when he was at Livermore Labs; he knew it could be done for less than one tenth of current costs. Among key economies, the Enkido network would contain no SONET, no ATM, no regenerators, no optoelectronics. Running the Internet protocol (IP) directly on WDM—and using a variety of relatively cheap innovations—Shafei obviated the bulk of existing costs.

For WDM transmission and transport, the young Egyptian-American turned to Corvis, a group of renegade engineers led by Ciena founder David Huber, who had mastered the problem of sending hundreds of lambdas across three thousand miles of fiber each at 10 gigabits per second, without regeneration. To adapt Lucent's sonstandard lambadas to the network, Shafei used gear from Avanex, led by Simon Cao, a Taiwanese with some 36 telecom patents issued or pending, and inventor of PowerMux technology. PowerMux will adjust the wavelength to the particular bit stream dynamically, shifting from 2.8 gigabit video streams to 622 megabit data on the fly, and accommodating as many as 2,000 wavelengths on one fiber.

For passive lambda switches, Shafei turned to a new company called Chorum in Dallas, Texas, headed by Scott Grout, formerly a leading optical engineer at Lucent. The long-haul reaches of the network would be all-optical. Only for the access points in large cities would Enkido use electronic technology in the net.

These arrangements alone would suffice profitably to serve Enkido's first customers, the HDTV broadcasters Fox and NBC and the government agencies, DARPA and NASA. Most current video transport runs through satellites with transponders that cannot operate above 45 megabits per second. High resolution or lossless video distribution takes gigabits per second and has virtually nowhere else to go but to Enkido. Shafei's backbone also could capture a large share of the increasing market for radiographic and other medical images which, by law, cannot be compressed. Enkido also is arriving just in time to win a significant portion of the coming market in 40 megabyte photographic images from Carver Mead's new professional Foveon camera.

Shafei's most potent disruption, however, was his ingenious solution for the long-baffling problem of last-mile broadband. In this effort, he turned to companies with fiber networks reaching throughout the country but with no stake at all in the existing telecommunications system, namely power utilities. In exchange for 20 percent of the revenues, Enkido leased a total of some 800 miles of local loop fiber from Pacific Gas & Electric in Los Angeles and in Silicon Valley and Los Angeles, ConEd in Silicon Alley in New York, and in Washington, D.C. All these companies run fiber to power substations serving some twelve homes apiece. From the substations, the links are powerlines.

This is only the first step, designed to develop a market for the superior solution: fiber to the home. Economically justifiable for telemetry alone, all Enkido's powerline partners promise to install fiber to homes and businesses as soon as the market is demonstrated.

Beyond Enkido's cheap optics, Shafei also envisages radically cheaper billing. Existing telcos spend between 30 and 40 percent of their revenue in expensive billing systems. With a radically simpler pricing scheme involving bits and bandwidth alone, Enkido can make dual use of its existing network management software from IBM's Tivoli and HP's Open View. Enkido expects to spend less than one tenth as much collecting payments for its services.

In Enkido, Shafei is demonstrating that the complexities of telecom can give way to vastly simpler and cheaper optical systems. This does not necessarily mean that Shafei will prevail. But his visions and ventures have shown the way.

The old players offer the song of the siren. They are offering guaranteed "quality of service"—low latency, low jitter, and assured rates of committed bandwidth—through "policy based networking," prioritization, time-division muxing, rate shaping, flow control, Asynchronous Transfer Mode, flow-based queuing, weighted fair queuing, reservation protocol (RSVP), and other fashionable buzzes which are even more complex than they sound. In effect, these companies are supplying a different network for each kind of traffic.

Their algorithms theoretically enable you to provide guaranteed channels for full-motion video, CD-quality voice, super-bursty data, real-time transactions, secure financial flows, and palpable three-dimensional holographic kisses over the net, and charge differently for each one.

It won't happen. Everyone wants to charge different customers differentially for different services. Everyone wants guarantees. Everyone wants to escape simple and flat pricing. Forget it.

Needless to say, big bandwidth is not a perfect substitute for big software, but it will do. It offers no particular quality of service, network control, or committed rates and latencies. Big bandwidth doesn't guarantee anything because it doesn't know what it contains. But it is advancing its cost effectiveness at least forty times faster than big software. In some ways, big software has been going backward. With a pace of advance forty times faster, you can make up for a lot of latency and jitter, misalignments and misfits and missed dates.

Today the ascendant technology is optics and the canonical abundance is bandwidth. Companies focused on jamming more and more information packets down a single-lane bit-stream—as if there were no bandwidth to spare—will lose to companies that waste bandwidth in order to build capacious multi-lane highways with each lane running well below capacity.

For lanes, think wavelengths or lambdas. The networks of the future will rarely need lightpaths that can bear the Net traffic of entire cities on a single beam. What the new networks will soon require is millions of addressable wavelengths of colors of infrared light. Each one will constitute a potential circuit connection between one terminal and another, just as your telephone creates a circuit connection between one user and another.

This mandate is difficult for most experts in the industry to grasp because it seems to reverse every important advance in networking over the past three decades. The leaders of telephony came of age years ago in an environment of bandwidth abundance, at least when measured against the modest demands of voice. They wasted bandwidth as a matter of course. Most of the capacity of a telephone network lay fallow more than 95 percent of the time as people used their phones an average of 20 minutes a day. In a world of bandwidth abundance, circuit switching—connecting the two parties over a line devoted entirely to their call—made sense.

As the Internet rose and data became dominant, however, computers remained on line for many hours at a time. From the Internet, telecom engineers laboriously learned the rules of packet switching, cutting up every message into many packets each bearing a separate address. While a circuit-switched phone network sets up the call in hundreds of milliseconds, a packet-switched network functions like a multi-megahertz post office. The envelopes are switched not in days or even milliseconds but in microseconds.

Finally, telephone and data network engineers have definitively learned the superiority of packet switching. Vinton Cerf of MCI WorldCom, the co-inventor of the Internet Packet protocol, TCP-IP, gives out an off-color T-shirt emblazoned triumphally: IP ON EVERYTHING. But if Cerf's company is to stay in the Telecosm, it will have to learn to restrict its IP processing to the appropriate places—on the edges of the network.

The new optics is turning the entire world of networks upside down once again. Wavelength division multiplexing many colors of light down a single fiber thread is ushering in a tide of fabulous bandwidth abundance. In a world of bandwidth abundance, bandwidth-wasting circuits become ideal once again. Rather than economizing on bandwidth by chopping everything up into packets and multiplexing them into time slots, the mandate is to waste bandwidth. As in the old telephone system, the best approach is circuits. In this case, the system software sets up wavelength circuits between terminals at the edge of the fiber network.

State-of-the-art fiber technology now permits some 2,000 wavelengths or lambda on a single fiber thread, adding up to 1,728,000 potential circuits separately addressable in a single fiber cable of 684 strands. If Shafei's Enkido dream is to be fulfilled, he must learn how to use this new resource of circuits, wasting the abundance of bandwidth created by his optical technologies and embracing the vision of Will Hicks: a lambda number for everyone in the world.

Shafei comprehends that he and the other gladiators of glass are engaged in a truly titanic struggle. When they are through, the old Departments of Posts and Telecommunications (PTTs), the TV broadcasting establishment, and the degraded culture of mesmerized masses will be swept away and the Great Pyramid will be gone.

Neither Shafei nor anyone else fully knows what will emerge in their stead, but they do not fear the future. Beyond the rainbow, Gilgamesh and Enkido found themselves on the other shore confronting the residual limits of human life, mortality, and light. A richer wisdom and a wider wealth would seem a boon worthy of the winning.

Chapter 16

Searching for a New Intel

Any party animals around here? Lurking longingly on the edge of the telecosm? Wistfully regretting mischances of yore—the Microsofts manqué, the Intels spurned, the Ciscos jilted? All while you awaited for a safe sell-off and a reasonable P/E, without any gut-churning bad news attached! Well, you're not too late—do we ever have a rabble-rousing, quantum-dancing millennial jubilee for you! You don't even have to mess with biotech. A new paradigm is emerging to rival the rocket that lifted Wintel into orbit.

Of course, we cannot be sure exactly which company will win. So we're off to a premonitory bash in February 1999 in Switzerland. Beginning on the alpine reaches of the Nasdaq, then moving on to the iridescent slopes of Davos, then effervescing in a mountainside hotel in the midst of an awesome blizzard, capitalist carousers are hailing the arrival of a potential new Intel of the telecosm. Its name is Uniphase.

Up here in the Alps, Uniphase has gathered most of its key backers and technologists. The nominal occasion is to celebrate a merger with Philips Optoelectronics in Einhoven. The cultures of money and optics happily converge, like the electrons in a laser. Buoyed by a P/E ratio well north of 60, all the Uniphase money is mingling nicely in phase, lighting up the room.

Dispersed through the party are more formal types, wearing the dark suits and ties that in California culture signify servants. They seem to be performing some facilitating role—dispensing wine and smiling warmly at their presumed betters. Their leader, a man named Volker Graf, would soon be attentively collecting information from guests on sizes and preferences in ski boots and bindings. Agile on the slopes and down the aisles of the bus to Davos, where Graf personally handed out everyone's lift ticket, these humble men turn out to be the Swiss scientists at the center of all the talk of a new Intel. They were simply extending old world hospitality to the boisterous Californian money.

The star of the firmament, however, is Josef Straus, a short, bearded gnome with a black beret, who headed a company called JDS-Fitel, Uniphase's leading rival in optic components. While Uniphase makes the active devices, JDS makes passives, the equally vital components that channel and modulate light without power. Rumors (which will prove correct) suggest that JDS will soon merge with Uniphase in a new optical powerhouse—a potential Intel of the telecosm.

• • •

The search for a new Intel is fraught with *son et lumiere.* A "new Intel" is obviously an analogy. But Jean-Baptiste Fourier, the French polymath and counsel to Napoleon, showed that analogy—essentially the recognition of similar paradigms in different fields—is among the most powerful modes of thought. Fourier's critical insight for communications—the convertibility of wave action from the time domain to the frequency domain, and back—emerged from a study of the movement of heat through a metal bar. "I value above all other tools the precious analogies," he said, "they open all the secrets of nature."

Today, callow interns in public relations at telephone companies speak in breathless tones about translating from the "frequency domain" to the "time domain." A wireless company recently sent me a tape announcing with trumpets a shattering breakthrough into an apparently "new dimension" in communications. Though intrigued by the concept of ultrawideband low power transmissions, I was crestfallen to discover that the new dimension was "time." The company called itself Time Domain. I thought I had already heard of time.

Then, citing as mere steps toward a new revelation the works of Copernicus, Newton, Maxwell, and Einstein, another vessel of magic called Silkroad declared at a Cipriani temple in Downtown Manhattan that it has revolutionized fiber optics by introducing Synchronized Refractive Communications. Wavelength Division Multiplexing (WDM), the current revolution in the field, thrusting optics far ahead of Moore's Law? Poof. "Obsolete and ineffective."

Behind the kimono, Silkroad turned out to have . . . more kimonos. They were fluttering with new solutions to Maxwell's equations, and they changed everything, incorporating . . . get this . . . the "time domain." It was possible to imagine that this was some practical joke. But as an explanation, chief scientist Jim Palmer sent me two books weighing as much as an Olympic standard shotput, and denser. Level Three is said to be interested.

This is an era of optics and optical illusions. With PR puffers Burston Mosteller at hand, luminiferous ethers surround us, undulating meretriciously in the light, among the fluted columns of Wall Street, fun house mirrors and exotic vapors, seeking a population inversion of photons, or money, whichever comes first. Literally thousands of companies are rushing toward the lights. Many do a bump and grind, lay some fiber or flash some lumens, and implicitly implore, "Buy me." But coherence is rare. For perspective, it is useful to return to Fourier.

The time domain is where we live, experiencing one thing after another, in a baffling onrush of sounds and images. The frequency domain—measured in the orderly Hertz of cycles per second—was rendered accessible by Fourier. Most communications systems work by using both these regimes.

The conceptual source for most of the spectroscopy and signal processing of the telecosm, Fourier transforms translate into intelligible signals every communications payload, from Wavelength Division Multiplexing of World Wide Web images to speech recognition of numbers over telephone lines. Fourier showed how any repetitive wave motion, however murky and irregular, from a modem screech over twisted pair to your voice, from the image of a hockey game to laser bursts of infrared light mixing hundreds of phone calls over fiber, could be translated into a set of regular sine wave frequencies. Assumed to be infinitely extensible, Fourier frequencies reach beyond time. Pure tones or colors identifiable by a spectroscope or producible by simple oscillators, Fourier frequencies are the key idiom of the telecosm, crucial carriers for all radio and optical communications.

Because of the Fourier phenomenon, any regular wave could be modulated or distorted with a message, combined with dozens of others in the time domain, sent over long distances, and then recovered, leaving the messages intact on their separate frequency domain carriers. Remove the regular frequencies, which are the carriers, and the residue is the message. Because, unknown to Fourier, photons are nearly massless, the messages can go two ways down the fiber without colliding.

When appraising the onrush of Internet valuations and volatility, think of an exponentially increasing radiance cascading toward the heavy concrete and steel structures of global commerce. Some of the existing infrastructure will survive in the palpable domain, unfolding in conventional time. The revenues of Federal Express, for example, are rising faster than Amazon's are; Barnes&Noble has become a new library and warehousing system; and Fedex may well clear more per book than either. But most of the transactions and interactions will move toward the highest reaches of the frequency domain.

The companies of the paradigm are mostly engaged in the process of moving the business of the world into the universal radiance of massless photons. There, the potential expands at the amazing pace of traffic doubling every hundred days or so for the last several years. Once you comprehend this picture in the time domain, and contemplate it seriously, there comes a moment of dazzling truth, and then, very often, a blind reach for the shares of any company that seems to have a hold or a handle on the incandescence.

In the new era, rather than using transistor switches and power to compensate for inadequate bandwidth, successful companies will race down the learning curve with ever more cost-effective broadband optical components.

Uniphase is a prime candidate to lead this race. Headquartered on Baypointe Parkway in San Jose, just minutes up Route 101 from the Olympus of Intel's Robert Noyce Building, even as late as 1998 the company was still obscure. Without making a major mark it had spent nearly two decades producing a motley assortment of lasers and other lightwave widgets. But through a shrewd marriage of Silicon Valley bravado and leading-edge photonic research from IBM, United Technologies, Philips of Holland, and Southhampton University, Uniphase has become an ascendant force in the surpassing millennial technology, wavelength-division multiplexing. It is a golden key to the telecosm, and Uniphase is just in time for it.

To become an Intel is no small challenge. In its three-decade history, the company founded by Noyce and Gordon Moore has grown by an extraordinary average of more than 25 percent annually. Intel's 1999 revenue—some $30 billion—was nearly double that of its nearest competitor, NEC. What could turn an obscure manufacturer of optical devices into such a behemoth? A transformation of the calculus of critical abundances and scarcities, which shapes economic growth and opportunity—a new paradigm. So comparisons must begin at the last such dawn, more than a quarter century ago.

On November 15, 1971, when Intel announced "a new age of integrated electronics," the embryonic giant was not the sole producer of any of the key components of what would become the personal computer. A team at Texas Instruments under Gary Boone had already made the first 8-bit microprocessor. A now forgotten Advanced Memory Systems was first out of the gate with DRAM, exploiting an invention by IBM's Robert Dennard. The electrically programmable read-only memory (first known as the EPROM, now upgraded to the "flash") was invented almost simultaneously at Intel by Dov Frohman and by a freelancer in Japan named Shumpei Yamazaki. The silicon gate process for reliably manufacturing chips had been created at Fairchild by Federico Faggin shortly before he joined the Fairchild defectors at Intel. But only Intel put it all together, in a single company that would go on to spearhead the growth of the industry—and the world economy right behind it.

At the heart of Intel's leadership was one felicitous technology of metal oxide silicon that inspired Intel's engineers to speak in an idiom of miracles. Moore himself famously noted that the key substances used on a microchip—sand, oxygen, and aluminum—were the three most common elements in the earth's crust. Equally uncanny was the discovery that, uniquely among elements, the silicon substrate of the chip could be both chemically protected during manufacture and electrically insulated during use by its own easily grown oxide. And most astonishing of all, in silicon the tradeoff between speed and heat generation improved drastically as size diminished; smaller transistors ran faster, cooler, and cheaper. *The less the space the more the room.* Sand piled on sand in a providential confluence, endowing Intel as the heavyweight champion of the microcosm.

Lying at the foundation of fiber optics is a similar set of felicities. The purer the silica, the more transparent—and therefore, effective at carrying signals—it is. Single-mode fiber has shown the power of ever thinner, ever more capable glass threads. In the telecosm as in the microcosm: *The smaller the space, the more the room.*

Wavelength-division multiplexing, WDM, brings this law to a luminous climax. What it requires in practice is daunting: keeping exquisitely defined and unchanging colors of light pure and separate, down thousands of miles of fiber and through scores of amplifiers, splitters, couplers, and other devices. At every point, physical deviations and delays can result in interference or dispersion.

In a glass prism, the crystalline structure of the glass forms a natural "grating" that separates a mixture of wavelengths into its component colors through diffraction and interference. In wavelength-division multiplexing, thousands of exquisitely laid-down lines per millimeter do the same thing. Made on glass or "speculum metals" (copper-tin hybrids), artificial gratings can perform all the basic functions of an optical network, selecting, diverting, and focusing light.

Gratings of all kinds require a complex and precise etching process. Used as external devices, they also require fusing or splicing the tiny fiber, raising a host of other technical problems.

Key to the emergence of modern wavelength-division multiplexing was a breakthrough that still mystifies its leading practitioners. Scientists in the late 1990s learned how to inscribe dense arrays of optical stripes, each with a different refractive index, *within the very core of the fiber itself.* As many as ten thousand grating regions could be imprinted in a few centimeters of glass, each governing a specific wavelength. It is as if Intel had figured out how to etch a chip *after* it was fully formed in its plastic package. Other in-fiber advances followed, allowing engineers to diffuse amplifiers and dispersion compensators through miles of fiber. And with that, a new field of in-fiber processing opened to view, including the possibility of productivity advances in silica as fruitful as Moore's law has been in silicon. An Intel of the telecosm was suddenly possible to contemplate—a company that manufactures the key components of the telecosm as Intel does for the microcosm.

The first tentative steps took place in 1978 in the form of a laboratory mishap. Scientists at the Canadian Optical Communications Research Center near Toronto ruined, "bleached," a stretch of fiber by shining a blue-green laser down its length. It would no longer pass light. Normally such an event would not lead further.

One of the researchers, Ken Hill, was intrigued. The laser in essence had transformed the fiber into a mirror that reflected all the light that entered it. Perhaps a more controlled operation of the laser might change the fiber into only a partial mirror that would reflect only some wavelengths of light. That would make it a filter. Following this line of reasoning, Hill tried exposing the fiber core to two opposite beams of mutually interfering laser light. Later research showed that the grating strength increased by the square of the laser's imprinting (i.e., writing) power.

For half a decade, this result remained an impractical lab anomaly. In 1983, however, there was a breakthrough. Eli Snitzer, then in his fifties, and two other researchers at the United Technologies Research Center in East Hartford, Connecticut, announced a far more efficient "Method for Impressing Gratings Within Fiber Optics" as they titled their paper. Patented in 1988, it opened the way to a new era of integrated fiber optics. Rather than focusing the light on the end of the fiber, they aimed the two beams from the side. They radiated the fiber with germanium ions, then blasted it with a laser. The UTC researchers were invoking a once exotic, now increasingly useful process—called stimulated Raman shifting—that uses phonons (sound vibrations) to amplify or change a frequency.

Using two beams of angled ultraviolet light, the researchers imprinted a permanent holographic image in the silica core. They also found they could adjust the grating features by "shaping and tilting the writing pattern through control of the included angle and divergence of the beams." In other words, engineers could use the technique to diffract, reflect, and refract wavelengths as they desired. The fiber artists were developing a palette with which they would eventually be able to paint masterworks in silica thread.

With the recommendation of his former colleague Will Hicks,

Snitzer in January 1984 left United Technologies to head up a major optical program at Polaroid, then one of America's leading high-tech companies. Seeking a replacement, UT turned to MIT's famed Lincoln Laboratories. Lincoln was the source of a wide array of innovations in computing and communications, from the Whirlwind minicomputer that led to the foundation of Digital Equipment Corp. to Paul Green's rake receiver, crucial to cutting-edge cellular phone technology. At the time in 1984, Lincoln's group leader in optoelectronic devices was Fred Leonberger. Although he did not command the scores of historic patents and papers of Snitzer (he was twenty years younger), he had worked on a number of key technologies for photonic networking.

At UT, Leonberger went to work with Snitzer's old colleagues Glen Morey and Gerry Meltz, and quickly developed an entire line of fiber-grating products, assembling seventeen key patents. In one solid commercial hit, he developed an optical gyro that used interference effects to stabilize cruise missiles.

By 1994 Victor Mizrahi, then a Bell Labs gratings expert, now at Ciena, could sum up the promise: "Fiber-phase gratings have moved from a laboratory curiosity to the brink of implementation in optical communications systems. In a few years, it may be as difficult to think of fiber-optic systems without fiber gratings as it is to think of traditional optics without the familiar laboratory mirror."

But "on the brink" is not over it, any more than a carefully controlled lab can substitute for the rough-and-ready telecom field. Before wavelength-division multiplexing could be deployed, its proponents had to prove that all their exotic new components not only worked, but would work under conditions that would make most microcosmic technologists blanch.

The weak link, as Snitzer had seen from the outset, was the laser in erbium-doped amplifiers. Always operating at high power, these so-called pump lasers are as different from the $4 types used in printers and CD players as a gigahertz server chip is from the beater in your old PC. It must operate around the clock, everywhere from the ocean floor to underground trenches, with a mean time between failures—MTBF—of a million hours or a hundred years. But at the time they were failing constantly, mysteriously, and catastrophically.

Back in the early seventies, Andrew Grove was the then unknown

Intel engineer who led the effort to solve a similar mystery in semiconductors. Chip manufacture was plagued by quality control problems. First thought to be caused by cosmic rays—the suggested solution was entrenching the chip fabs underground—the catastrophic breakdowns turned out to be impurities in the silicon, an effect of sodium accumulations in the diffusion furnaces. Cleaning the furnaces was the answer—not laboratories built in bedrock like bomb shelters.

This time, the then unknown engineer leading the team that solved the problem was another European working for an American company, Volker Graf. And once again the problem was impurities: Oxygen atoms were infiltrating the laser and causing rapid oxidation, also known as rust. Graf and his team invented a method for blocking the oxidation and eliminated the problem. Then, facing corporate cutbacks, he persuaded his bosses at IBM to spin out a separate company, IBM Laser Enterprise. By the mid-1990s, Graf was running a profitable $15-million-a-year business producing reliable pump lasers, including a 70 percent share of the soon-to-explode telecom market.

By the mid-1990s, three of the world's technology giants, United Technology, IBM, and Philips, had succeeded, by dint of heavy investments over decades, in developing key components for WDM. Collectively they commanded more than a hundred Ph.D.s in physics. In Zurich, Einhoven, and Bloomfield, Connecticut, they operated the leading edge factories in the field. And yet, at the very moment when the WDM component market was about to explode into a $6-billion business, these companies *were giving up fiber optics.* As Clayton Christensen would show in his book, *The Innovator's Dilemma,* this decision was probably right. Fiber optics was not part of the core business of any of these companies. Separately, their assets could have fueled fast growing and managerially demanding optical components divisions for at each of the leviathans without making a noticeable contribution to their total revenues. Put together, however, the divisions of the three companies would make a new high technology powerhouse, a spearhead for the new optical paradigm, maybe even a new Intel.

To Uniphase, which began the 1990s deep in the old technology of gas lasers, the new paradigm seemed to offer nothing but turmoil and risk. Based in faraway Silicon Valley, the company was a niche manufacturer of helium neon gas lasers, founded by Dale Crane, an introverted engineer and veteran of the flagship 1970s laser startup, Spectra

Physics. Spectra, the most successful company in the laser business, made bulky but high-resolution helium neon devices for supermarket bar code scanners, the sort that allegedly bedazzled George Bush. A shrewd designer, Crane envisaged miniaturizing gas lasers, and moving them out from their fixed platform under supermarket counters. But seduced by lush defense contracts, his bosses saw no commercial demand for a smaller and, presumably, inferior gas laser device.

In 1979, with $300,000 of his own money, Crane bolted Spectra Physics and retreated to a garage in Menlo Park to found Uniphase. Within a year he developed a more compact helium neon gas laser, hand-holdable by a clerk at the counter or down the aisle in a warehouse. Symbol Technology, a manufacturer of portable devices, started buying twenty thousand Uniphase gas lasers a month. By the mid-1980s, the company was generating some $30 million in revenues and far outpacing its parent in unit sales of helium neon devices.

Helium neon lasers emit a pure infrared beam in a very narrow frequency band, enabling them to read dense bar codes from the briefest swipe across the lens. You don't have to be a campaigning president to be amazed at this. But with demand for the devices close to saturation and competition on the way, Crane wanted to move up market and up spectrum. He engineered a realiable argon gas laser that produced blue light. Long a grail of the industry, blue lasers could carry Uniphase into biotech, medical, and semiconductor instrument markets, which needed short, visible wavelengths to detect submicron phenomena—particles on a wafer surface or microorganisms in a petri dish.

Crane went on to develop an array of more efficient CO^2 and other gas lasers, satisfying a variety of niche markets. His technical virtuosity and continued profits, even at very slow rates of growth, worked well enough for him and his customers. In his view, he had created an eminently successful private company, a technical marvel in its field, producing the world's best lasers. What more could anyone want?

By the early 1990s, however, Uniphase's board of directors, led by an independent venture capitalist named Milton Chang, was restive. A private company, no matter how stable, is a closed cavity; it is hard either to get money in for expansion or out for shareholders. Without this systole and diastole of enterprise, growth tends to languish. Chang persuaded the board to seek a new CEO, who could spruce the company up for an IPO. A headhunter contacted Kevin Kalkoven, then a turnaround specialist for two venture capital firms, Oak and Merrill-Pickard. A laser enthusiast by osmosis, thanks to two friends with gee-whiz interests in holography, Kalkoven was intrigued. But first the Uniphase board wanted him to see an industrial shrink for testing.

• • •

For Kalkoven, the telecosm would arrive with all the abrupt pain, the rupture, and heavy breathing of a major life crisis—such as the death of a father, loss of a country, marriage, childbirth, divorce. He had undergone them all.

When he was age two, in 1944, his father had returned home to Adelaide from the Burmese front, with a suppurating wound and a tubercular lung. A few months later he died, leaving Kevin an only child in a one-parent family, living on a military pension. By the time he was eight, his mother put him to work weeding gardens, running errands, delivering newspapers. Child labor was his redemption and his crucible, as it is for many entrepreneurs. When he was twelve, even his mother seemed to betray him, marrying again. The buffeted youth entered a rebellious adolescence on the streets of Adelaide. Just the sort of rocky childhood, perhaps, to prepare someone for leading a small company into the combat of a convulsive new era in telecommunications. Or perhaps not.

When the call came from Uniphase, Kalkoven was bearded, paunchy, and in his fifties, with a penchant for flying small planes and scuba diving. He had the self-indulgent air of a man who had yet to discover his life's true calling and who liked to have fun, with an Australian accent. Divorced, with two daughters, he was living with a longtime companion Sally, whom he had met during a stint with Comshare in the U.K. A COBOL software programmer with IBM in the early 1960s, he had later moved to Michigan and then to Silicon Valley, where he had conducted modestly successful turnarounds at a number of enterprise software firms. It was an appropriate résumé for someone seeking to run a company addressing, say, the year 2000 glitch. Nothing suggested any lasing candle-power—any obvious bent for transforming Uniphase into the Intel of the telecosm. But perhaps he could get the investors some return on their long engagement with Dale Crane and his ventures in gas lasers.

But first there was this matter of a cautious board and their hired inquisitor. In fact, the psychological test was at least partly a cover that allowed the psychologist, Claudio Serifini, to lurk in the corridors of the company. The board needed the shrink to help ease Crane through the tensions of getting rich through an IPO he devoutly opposed, a company reorientation he deemed entirely unneeded, and the hiring of a new CEO who knew little about lasers. By then the owner of less than 40 percent of the company, Crane made it clear that he would not voluntarily relinquish control to unknown outside investors or catered executive talent. The board feared a blowup.

So *everyone* took the tests, including senior executives and applicants for the CEO position, who were gauged for emotional stability, leadership, and career aptitudes. The board knew that the firm could not attract a superstar executive. But it might find someone who was experienced, reliable, and capable of handling an awkward transition.

As Kalkoven remembers, the psychological test put him far on the "intuitive" side of the scale, which in Silicon Valley is a polite way of saying: "Can't do math, but too imaginative to go to law school." In addition, he couldn't draw or dunk a basketball, and was too greedy to want to teach. If the truth be known, he also suffered a vision defect, failing—along with nearly all of his Silicon Valley contemporaries—to see the coming Internet tsunami. In his case, he professed to find the computer industry "too slow" for his tastes.

Nothing to do but make him a CEO of a small woebegone optics company, with the apt name of Uniphase (a reference, incidentally, to the key property of laser light: it is all in one phase). Joining the company in January 1992, Kalkoven faced the challenge of bringing the company public, in defiance of existing management, before anyone discovered that the market was mostly saturated and that the laser upside, if any, was in semiconductor devices, which Uniphase didn't make. And there was a pesky patent suit from Spectra Physics, with total lawyers' bills running toward $2 million and little prospect of victory. Lose the suit, and there went the argon laser business. The board saw an IPO as a way out for faint-hearted investors and a money-raiser for new projects. But in his first attempts to push it, Kalkoven made virtually no progress, even with close friends in the banking community.

It was clear that the company needed to improve its story. Sometimes, reasoned Kalkoven, you can turn around a company by dominating a declining market, from which the stars are defecting. So Kalkoven decided to add to Uniphase's obsolescent product line still more gas lasers. The low point in this effort came when he and Dan Pettit, the company's able chief financial officer, flew to Munich to make a bid for Siemens' gas laser operation. As Pettit recalls, the Siemens division commanded revenues of some $15 million a year, with low margins and sluggish growth. Pettit calculated the discounted present value of the cash flow at $4 million; he and Kalkoven bid that amount. The Siemens representatives sneered and indignantly left the room.

Their trip to Munich in a shambles, and a meeting with New York bankers scheduled only thirty-five hours hence, the two men stood in the conference room and looked at each other grimly. Then Kalkoven brightened: "Hey, let's go to Paris," he said, mentioning a friend named O'Brian who "knew some bankers." It seemed a good idea. Meeting O'Brian at a luxury restaurant, Kalkoven began drinking heavily, offering loud bets on the professions of other customers, and careening across the floor to ask what they did for a living. Not for the first time, Pettit began to wonder about his new CEO.

Hastening from the restaurant, our three heroes then left to search for investment bankers and underwriters on the Place Pigalle, among the night clubs, bars, sex shops, and offers of "live girls." Although they found no willing investors, a barker at one of the brothels came out and steered Kalkoven toward the door, then began pulling him by one of his arms. Although Kalkoven "had a silly smile on his face," Pettit shrewdly guessed that any bankers inside might not be in the mood to consider laser technology. He grabbed Kalkoven's other arm, and pulling desperately, saved his CEO from an unknown fate within.

After going to bed at 4:00 A.M., they awoke the next morning and set out for their meeting at an investment bank in Manhattan. The time zones were on their side, but they missed their plane to New York, which was probably just as well. Kalkoven recalls a later session with potential underwriters in Massachusetts, who were riding the crest of a legendarily successful offering, Boston Chicken. He could not get their serious attention until he spontaneously began touting a Uniphase "technology" for using lasers to sear chicken breasts. The company's prospects as a cuisinary laser equipment firm dissolved when he burst out laughing.

Kalkoven spent the next two years wrestling with the lack of profits then endemic in the whole laser industry. Lasers had a reputation as a scientific breakthrough, bringing an exotic quantum phenomenon into visible incandescence. They thrived in the press as a bauble beloved of all technology buffs, holographic dreamers, and light show luminaries. But nothing but disappointment met the early hopes evoked by Gordon Gould and others for using them to zap missiles from the sky or solve the oil crisis through fusion energy. There were too many other ways to perform these functions and lasers could never surmount Peter Drucker's requirement that to displace an old technology the challenger must be ten times better.

At first, Kalkoven tried to move Uniphase toward semiconductors. Hiring a materials scientist named Bruce Wooster, he launched a subsidiary called Ultrapointe, to use Uniphase's argon blue laser in chipmaking equipment. Detecting and analyzing submicron defects in three dimensions, the Ultrapoint device—"an MRI (magnetic resonance imager) for chips," as Kalkoven describes it—has been a solid success. Though imparting scant long-term strategic advantage, it served the purpose of increasing Uniphase's attractiveness to investment bankers: the firm could be presented as a solid laser company with an exciting breakout product in the semiconductor field. Art Samberg, of Dawson-Samberg, a semiconductor capital equipment enthusiast, was among the first to tout Uniphase. Mel Lavin of Unterberg-Towbin followed, along with Phil Lamoureaux, who had been steered into the Intel IPO by his friend, Gordon Moore. With Unterberg-Towbin leading, the firm went public in November 1993, at a valuation of $30 million. Winning IPO bingo with Pettit as promised, Kalkoven collected $11 million in new funds for the company. But he was still groping for a way to transform Uniphase into an enduring winner.

The turning point came early the next year, when MCI installed a fifteen-hundred-mile fiber-optic line between Sacramento and Chicago with no optoelectronic repeaters. Although the system used only single-wavelength technology, it was the first long-distance demonstration of all-optical networks and it made Kalkoven an instant true believer. "Lights turned on," he says, and sirens wailed in his mind. To Kalkoven, the MCI feat "showed that WDM was feasible, even inevitable."

The simple key to MCI's "miracle" was the erbium-doped fiber amplifier that Snitzer had conceived so many years before. In a moment of true luminosity, Kalkoven decided to focus Uniphase on acquiring all the tools and competencies needed for the essential components of optical networking. Since gas lasers, cumbersome and power hungry, would not work with this technology, his decision meant a transformation of the company from gas to semiconductor lasers—a turn from the old paradigm to the new.

It was a conversion experience. Kalkoven turned himself into a paragon of the telecosm, ready to defy all obstacles to his new vision. He was revitalized physically and psychologically. The hedonistic Australian reemerged as a reasonable facsimile—tough, smart, and usefully paranoid—of an entrepreneurial Andy Grove.

Bruce Wooster of Ultrapointe and his crew of gas laser experts and engineers exploded. Declaring Kalkoven "crazy," Dale Crane sold his shares and moved to Nevada with a new young wife to beget four children. With $5 million in hand, Crane at that point seemed the only financial winner of the new Uniphase era.

But Kalkoven remained unshaken. Wavelength-division multiplexing would be the new company grail. Consulting with board member William Bridges, an optical physicist from Caltech, Kalkoven set his first acquisition target: the United Technologies Photonics research division in Bloomfield Connecticut, whose chief product was Fred Leonberger's optical gyroscope for cruise missiles.

With military spending in free fall, United Technologies by 1994 was trying to unload its photonic offspring. Leonberger's gyro worked by sending pulsed light in phase two ways around a tiny circle and detecting skews in a missile's position from interference where the two light paths overlapped. But the gyro's inventor, still UTP's chief scientist, had also been working on a promising potential commercial application: electrically pulsed, a lithium-niobate gyro could modulate a laser beam, with the meticulous accuracy needed for WDM. Until that time, most fiber systems used internal modulation, turning the laser off and on at the appropriate rate—millions or billions of times a second. But as speeds went up into the gigahertz ranges and the number of WDM wavelengths jumped from one to four and beyond, Leonberger believed that direct modulation would eventually break down. And his rugged, reliable, low-power gyros would be the perfect replacement.

Kalkoven, his intuition in play, was sold, and he and Pettit negotiated an $8 million deal with UTP's corporate masters. But just as the contract was about to be signed at the Morrison & Forrester law offices in Palo Alto, Pettit received a phone call from Paul Suchowski, manager of UTP's modulator product line. Himself an engineer and inventor, Suchowski bluntly announced that UTP management, from Leonberger on down, had resigned. They wanted to go with Ortel, the mid-1990's fiber-optic star.

A longtime rival of Uniphase in lasers, Sunnyvale-based Ortel commanded a potpourri of state-of-the-art devices, including sophisticated gear for the cable industry and satellite communications. Ortel offered UTP's scientists the chance to develop exotic new technologies. But not incidentally, it was also offering them a chance to get rich, through generous stock options (at the time the Ortel shares were far outperforming Uniphase's). Suchowski declared there was no turning back.

United Technology headquarters asked whether Kalkoven and Pettit still wanted to make the deal. On the one hand, the two still wanted the patents, the technology, and the facilities. But without the intellectuals who made it, UTP's intellectual property was nearly worthless. Kalkoven and Pettit briefly conferred, gulped, and decided to go for it. They would somehow cope with the personnel crisis. If the scientists and managers wanted to go elsewhere, too bad for them. Kalkoven and Pettit signed the contract. Then they called Leonberger and told him that they had bought his company. After two weeks, he and most of the rest of the UTP team returned—now to Uniphase. Suchowski held out longer, but finally gave in. Kalkoven and Pettit had won.

Uniphase now had command of its first crucial component for WDM—a rock-solid external modulator that could operate at gigahertz frequencies. Kalkoven eventually invested his entire IPO war chest in the project, including $3 million for a new manufacturing plant in Bloomfield. All these capabilities—and Leonberger himself—had briefly seemed destined for Ortel. But Uniphase invoked the magic of bold action, fueled by the vision of an entirely new paradigm, advancing at a pace too fast for faint hearts or haggling. Ever since that fateful day in 1994, Uniphase stock has been ascendant and Ortel in the doldrums.

Beginning with the modulators from UTP, Kalkoven would assemble technologies to supply every crucial component for these revolutionary new optical networks, from EDFA pumps to fiber gratings to packaging and assembly. Most importantly, Uniphase now included many of the industry's leading scientists, including several hundred top optical physicists and engineers.

But Uniphase's pivotal acquisitions did not come until 1997, when Kalkoven, Leonberger, and Pettit made a bold raid on IBM Research Labs in Zurich. For $45 million, they bought Graf and his entire team at Ruschlikon, as well as their patents. Then in April 1998, Uniphase acquired Philips Optoelectronics group, in a complex stock deal rumored to be worth close to $150 million. Combining the photonic orphans of three paradigm-blind giants with their own snowballing entrepreneurial verve, Uniphase became the major component supplier to the explosive WDM equipment market.

But there was one piece of the paradigm puzzle that Uniphase conspicuously lacked: fiber gratings. Although many of the field's key discoveries had come from United Technologies, the company ultimately shuttered its gratings operations. Lucent, the telecom equipment leader, standardized on a different technology of semiconductor arrays for filtering WDM frequencies which seemed to make UTPs obsolete.

Meanwhile, Fred Leonberger had been meeting with a team of engineers at a tiny new WDM equipment maker called Ciena. Growing at a rate of over 100 percent per year, serving Sprint, WorldCom, Cable & Wireless, and the Japanese, Ciena quickly became Uniphase's largest customer for lithium-niobate modulators. In 1997, Ciena's astounding growth would fuel an IPO that valued the company at $3.44 billion, the largest stock sale ever by a startup.

It was Ciena's Wall Street breakthrough that enabled Kalkoven to raise the funds to fulfill his own dream—$120 million more in the next two years. But to gain an edge in the market, Ciena had also been forced to master the gratings Uniphase couldn't supply: "It was the only area where we decided to integrate vertically," says Pat Nettles, the Caltech Ph.D. who today leads Ciena.

As a result, under the guidance of former Bell Labs engineer Victor Mizrahi, Ciena has become preeminent in the field. Fiber gratings have propelled the company's WDM breakthroughs, first to 16, then 40 and eventually 120 wavelengths of light down a single fiber thread. The effort required ever more precise control. Most of the industry initially believed that 100 gigahertz of separation between bit streams was necessary to prevent the blurring of colors together (interference). But

Ciena has used its gratings expertise to create first 50 and more recently 25 gigahertz separations. And Mizrahi thinks this is just the beginning of a move toward Hicks's vision of as many as a thousand usable wavelengths per strand—each capable of using today's technologies of carrying between 2.4 and 10 billion bits per second.

As the would-be Intel of the telecosm, supplying key components to fiber-optic equipment manufacturers such as Ciena, Lucent, and Alcatel, Uniphase couldn't depend on one of its *customers* for so essential an ingredient. Kalkoven resolved to go for his own gratings—and as it happened, he knew just where to go. Using American Airlines Advantage miles like a good entrepreneur, he left California in early 1997 for Sydney, Australia.

Next to the airport there in the faraway Antipodes lurked one of the leading figures in the evolution of the industry, the tall, bearded coinventor of the second generation of erbium-doped amplifiers, Simon Poole. Less well known, but ultimately more important were his contributions to the development of robust fiber gratings. In 1993, Poole had left Southhampton University and British Telecom to return to his home in Sydney and launch a new company, Indx, to manufacture holographic gratings and other in-fiber devices. Using technology spun out of the Australian Photonics Cooperative Research Center—an optical consortium of government and university scientists—Poole's company had a proprietary process it was using to manufacture grating products for Asian customers such as Fujitsu.

Poole's central achievement was a technique that enabled engineers to overwrite a number of gratings with different refraction indices on top of one another, in the same section of fiber. Another feat of integrated optics, it enabled inclusion within the fiber core of a variety of WDM functions now performed externally. At the time, both Leonberger at Uniphase and the Philips team in Holland were moving to integrate as many as twenty laser functions on exotic indium-phosphide chips. If Kalkoven could link their techniques with Poole's grating innovations, the new Uniphase could truly fuse.

Freely dropping the names—Fred Leonberger and Volker Graf—and resonating with Poole's Australian origins, Kalkoven met with his countryman over beers at the Sydney Opera House. By the end of the evening, Poole had agreed to sell Indx to Uniphase for $9 million. With expertise in add-drop multiplexers and dispersion compensation modules as well as gratings, Indx would spearhead Uniphase's drive to bring the all-optical dream into metropolitan and enterprise networks. Nothing would be so important to making Uniphase a real Intel.

Microcosmic Intel began with the goal of manufacturing memory chips for mainframe computers, but it did not achieve incandescence until it began manufacturing the key components of the personal computer. Similarly, networks cannot become truly all-optical until wavelength-division multiplexing descends from the summits of continental trunks and undersea conduits down into the proliferating ramifying realms of small networks. The true target of Uniphase must be to go beyond its role as chief supplier to telecom equipment firms such as Nortel and Ciena. To fulfill its destiny, Uniphase must also become the leading supplier to computer networking companies such as Cisco and 3Com, and perhaps, as time passes, even for Intel itself.

Nothing will advance that cause so rapidly as taking costly and slow electronic switches, routers, bridges, and multiplexers, loaded with software and circuitry, and replacing it all with simple optical components. To achieve this goal, Kalkoven would have to join his worldbeating active devices, such as high-powered lasers, with passive devices that could split or couple light through a neighborhood or a LAN. Shortly after the company's 1999 visit to Davos, Kalkoven officially announced the merger of his company with JDS-Fitel, the leading manufacturer of passives and a stock market star as meteoric as Uniphase. Partly negotiated between Straus and Kalkoven in an upstairs room at the Gilder-*Forbes* Telecosm conference at the Inn at Squaw Creek near Lake Tahoe in September 1998, the new JDS-Uniphase commands all the tools to shape the future of communications as Intel shaped the firmament of personal computing.

To consummate the Uniphase dream, Kalkoven and Straus, his successor as CEO, will have to swim down market, becoming a vendor of the key devices for networks reaching into every home and small office—linking the network computers of the future to the crystal cathedrals of an all-optical web.

Crucial to this new vision will be the creation of cheap optical chips. Whether Uniphase buys them or buys a company that first produces them, the analogy to Intel leads to the possibility of optical integrated circuits resembling the silicon "ICs" that launched the microelectronics revolution. The goal is the creation of tiny optical chips that can perform many different functions on a conventional silicon substrate. Manufactured in the millions, such devices could be sold for a few dollars (or even a few score dollars) and incorporated into every personal computer or other Internet terminal. Beyond the successes of Intel and other semiconductor companies, an inspiration for this second vision is the CD laser, a complex quantum well device that is made by the billions and sold for $1.50 apiece. By comparison, Uniphase lasers for WDM cost hundreds or even thousands of dollars apiece.

Cheap optical integrated circuits could allow the movement of photonic communications down into the last mile of the network, into campus links, and even into local area networks (LANs). Conceivably these chips could bring broadband photonic communications into the backplanes of PCs themselves. In the end, gigabit networks capable of high resolution video, video teleconferencing, and complex visualizations and simulations would become as cheap and prevalent as megabit networks are today. All the electronic bottlenecks would shatter at once and floods of light would stream into the innermost reaches of your computer.

The promise is there. The in-fiber devices of the telecosmic vision have dramatically lower losses (by a factor of roughly thirty-five), easier coupling to fiber systems (being inscribed in fibers themselves), much lower sensitivity to temperature without active thermal control, immunity to polarization, and lower complexity. All-silicon optical chips from companies such as Bookham in Abingdon U.K. promise adequate performance and hugely lower costs.

In the *Innovator's Dilemma,* Clayton Christensen tells us that cheaper, inferior devices can prevail at a time of technology overshoot, when the industry is giving customers products that exceed the needs of the bulk of the market. At present, optics offers a condition of technology undershoot. At a time of explosive expansion of Internet traffic, the optics industry cannot supply either the bandwidth or the easy add-drop and switching functions demanded by the market. Therefore Bookham today poses no threat to JDS Uniphase.

As long as the critical measure of success in the industry is reducing toward zero the cost per transmitted bit per second, the Uniphase vision will prevail. But for penetration toward homes and LANs, the critical measure will not be cost per bit per second. It will be cost per terminal and cost per router port. While typical WDM products command markets of tens of thousands a year, Cisco currently sells some 60 thousand routers a month and PCs sell by the hundreds of millions. A cheap, effective silicon technology might well prevail in these lower end slots. The initial result will be to expand the markets for leading edge telecosmic devices by further expanding Internet traffic. But the history of such disruptive low end devices suggests that integrated silicon optics will ultimately dominate many up market slots as well. I believe that the producers of high end optical devices are premature in dismissing the Bookham threat.

Bookham's advances in microcosmic integration are genuinely impressive and are likely to find niche markets in coming years. A homogeneous single crystal material, silicon is the best known, most thoroughly tested, most widely manufactured, most microscopically manipulated substance on the face of the earth. Already, Bookham has introduced all silicon multiplexer/demultiplexer devices operating at 16 channels and effective receiver-transmitters with eight channels.

Bookham has also built a wafer fab using conventional capital gear from the semiconductor industry and hired 150 people, including Peter Ballantyne from Bell Labs and AT&T to head the engineering and Kevin Ford of National Semiconductor and AMD to run the wafer fab.

Leaving Britain at night headed for Holland in a British Airways Boeing 757, I was captivated by their plan. I relished the esthetics of new integrated circuitry that could attach every computer to the web photonicly at gigabits per second, ultimately for a few score dollars apiece, the price of an Ethernet NIC (network interface card). I looked down from the plane at the motorways surrounding London, scintillating webs of golden light, streaming with a heavy traffic of automobiles, sparkling like so many IP packets riding directly on the hard clad silicon loops and I let myself imagine that Bookham had done it, brought the light down every British street, linked every Oxfordshire Lane in endless trunks and branches, hubs and roundabouts, brocaded across the British nocturnal landscape like gilded and luminous lace in the moonlight.

On arrival on the continent, where I was to give a speech at Uniphase in Einhoven, however, the Bookham analogy began to break down, at least in part. The analogy between electronics and photonics is powerful but incomplete. In the paradigm, technological progress is driven by key abundances, defined by the plummeting price of an essential factor of production. In the industrial era, the plummeting price was physical force, beginning as horsepower, ultimately expressed in kilowatt hours. In the microcosm, the driving force of progress was the plummeting price of transistors, reducing the cost per stored or processed bit toward zero. In the telecosm, by contrast, the driving force of progress is the plummeting price not so much of storing or processing bits but of transmitting them. Explained Chris Harder, chief of R&D at Uniphase in Switzerland, "In semiconductors, you get more and more transistors on one chip and everything improves. But in telecom, you don't necessarily want more lasers on one chip. You want more light pulses emitted into the fiber. Integration may not help at all."

In this view, photonics and electronics are inexorably different. At the heart of integrated circuitry is the phenomenon of voltage fanout—a voltage divider. You can divide a voltage many times without reducing it. Thus millions of transistors can be activated across the surface of a chip without attenuating the voltage. But divide photonic energy and you attenuate it. Each time you split it you have less. Thus integration is more difficult.

In electronics, as the wires grow smaller in diameter their capacity shrinks because of rising capacitance. The diameter of the wires has to rise by a factor of ten as you move each step from the chip to the backplane bus to the coax cable of the network.

In photonics the smallest fiber cores carry the most information farthest. Single-mode fibers with channels less than 10 microns can bear scores of different colors of light. The process of "integration" occurs in the fiber rather than on the chip as in electronics. This integration of many colors on a single fiber multiplies by a potential factor of several hundred the capacity of optical networks and has launched a new era of bandwidth abundance on the backbones of the Internet.

This is the real, new integrated circuit, integrated not on silicon slivers but across mostly silicon continents and sea beds, a global seine of silica that can capture all the glittering profits in the Internet's global ball of radiance. JDS-Uniphase is the current leader in the pursuit of the Intel dream. But it cannot look back. It must look to the ascendancy of the light.

Chapter 17

The TeraBeam Era

It is late February 2000 in Seattle and I am gazing through a dirty window in a small office building off Mercer Street on Capitol Hill. This is the land of fog and microsmog, drizzle and mist. Would you believe that I have a sunburn from a waterfront run this morning? Would you believe that behind this smear of window I am currently basking in the full blaze of the telecosm streaming through at two gigabits per second?

Catching the blast of light as it comes through the window and turning it into an incandescence of swashbuckling Sean Connery images, video teleconferencing banter, and file transfer gigabit bursts is a pastel box the size of a flat panel computer display. It holds a series of ingenious inventions that reduce its cost to $150. Can you spare a dime for a multimedia gigabit per second?

The company I am visiting is TeraBeam and it consummates the telecosm, extending the magic of lightwave communications to the wireless domain. Moving all the way up Maxwell's rainbow to the edge of the domains of visible light, Terabeam's terabits are on their way toward a location near you. An ingenious new point-multipoint technology that can serve many users at once from one antenna, it closes the last mile gap between the wavelength division multiplexed floods of fiber and your computers. Point-multipoint means cheap shared bandwidth, with each transceiver at each hub serving scores of customers. It is the first approach that can operate in the tens of gigabits per second in the last mile without the hassles and delays of buying spectrum and rights of way, fighting for space on crowded roof tops, trenching fiber and digging up streets.

I have been looking for such a company for a decade. In the confidence I could find it I denounced the spectrum auctions for LMDS. But finally I gave up and decided that there would be room for a microwave era before wireless optics inherited the broadband mantle. But surprisingly enough, it turned out to be tediously difficult to shuffle microwaves across a city. Continual delays afflicted the point-multipoint gear needed to serve cheaply the ultimate target population of some 700 thousand US office buildings. Literally thousands of companies have been struggling with the problem—from Nortel and Lucent to PCOM and Netro. Teligent, Nextlink, Winstar, and others have been rolling out their systems one by one at T-1 (1.544 Mbps) or T-3 (45 Mbps) bandwidths, nearly all point-to-point, with primitive pairs of microwave radios. By the year 2000, the total number of buildings served by microwave wireless still stood in the low thousands. In 2000, Teligent was still adding wireline connections faster than wireless ones. Over at Craig McCaw's Nextlink when we ask how the radios are doing, they say, "Great, great, we have demand backed up over a year and a half for new installations."

When we breathlessly exclaim our relief that someone has made Point to Multipoint radios that work, our Nextlink friend corrects us: "We meant point to point installations."

"What about multipoint?" we ask, knowing that up spectrum wireless bandwidth is a much less attractive alternative to fiber if the network must be stitched together from point to point links.

"Well, they're trying very hard."

In a world in which all the crucial Telecosm technologies are rolling down hill, always beating out last month's bandwidth projections, up spectrum radio seems to be pushing a bandwidth boulder up hill. Our friend asserts his confidence that the $700 million Craig McCaw spent on LMDS spectrum won't be wasted, "because he is a man of vision." Perhaps he can sell it to AT&T.

I have been listening to the technology. Could it be trying to tell me something?

As TeraBeam founder Greg Amadon puts it, "For the $28 million that McCaw paid for LMDS spectrum in Seattle alone, we can build an entire 100 gigabit per second system." That means a cell with four sectors each handling 24 customers with one gigabit per second downstream links. To the Telecosm's crystal cathedral, TeraBeam adds oudoor lighting.

The first effect of the new wireless optics will be to disrupt the business plans of all the 24 gigahertz (Teligent), 28 gigahertz (Nextlink) and 38 gigahertz (Winstar) microwave companies. Less affected will be the vendors of spread spectrum fixed wireless service to residences and smaller offices and the MMDS (Multipoint Microwave Distribution Service) plans of WorldCom at 2.5 gigahertz. These longer wavelength systems may well find application in towns and rural areas. Depending on whether the system is in foggy Seattle or sunny Tucson, TeraBeam requires each point in the network to be within three to 10 kilometers of a hub.

But a technology so massively superior is chiefly a bearer of huge opportunities. The amazing bitstream pouring through this window on Mercer Street throbs with a High Definition Television stream (Sean Connery cavorting in *The Rock*), five big test files of a gigabyte apiece, and an online video teleconference with no perceptible lags, in which an engineer is responding to my questions about the broadband flood. (Among TeraBeam's inventions is a way to remove the half second delay from MPEG2 video transmissions, rendering them suitable for video teleconferencing).

Two kilometers away from Mercer Street, across Lake Union on Queen Anne's Hill, marked by a visible red light for tracking, is the source of the bitstream. It is an invisible infrared beam at a 1,550 nanometer wavelength. In the past, such a bitstream would demand a fiber optic link. But between us and the house on Queen Anne's Hill is nothing but light and air. It turns out that I was correct in speculating ten years ago that if you could send a 1,550 nanometer optical signal 1,000 kilometers down a fiber thread one tenth the width of a human hair, you might eventually be able to send the signal three kilometers without the thread.

Since the early 1990s, the fiber distances have lengthened to 3,000 kilometers, the colors of usable light have multiplied to hundreds, and erbium doped amplifiers have gained several watts of power. Now, after more limited successes by Canon, Jolt, and Lucent, TeraBeam's group of spectronic wizards in Seattle have mastered most of the challenges of wireless optics. Fiber insulates the electromagnetic signal in glass; TeraBeam insulates it in air. But it is the same stream of pulsed waves.

The story began in March 1997, when a photonic epiphany struck Greg Amadon. A serial entrepreneur with minor successes in cellular phones and optic devices, he was also working with engineer Richard Rallison, who had launched a LIDAR (laser radar) system used for detecting windshear at airports. Amadon found his attention wandering toward the window during an entrepreneurial luncheon in the Chairman's room at the Columbia Tower Club. Atop Seattle's tallest building, it was a site where the mind naturally tends to slip into a self-congratulatory haze of condescension toward the rest of the world spread out so far below.

Instead the twenty venturers were animatedly chattering about the Next Big Thing, which they assumed would be a belt borne wireless device that could discreetly divulge to others at a bar your availability, preferences, and technical specifications. Of course, if the specs were truly definitive, why go to the trouble of trundling down to the bar? Just send a virtual simulation of yourself over the net. Written in Java, it could be implemented on any operating system without disturbing you at all. Or it could be compiled by a computer aided manufacturing program and extruded in cellulose by one of those devices that delivers plastic models of specified designs. If you get a bite or a buyer, you can hold an e-bay auction.

Anyway, if your amusement is already wearing thin, you can understand why Amadon began ruminating on wireless optics. He found himself gazing out the window at the scene below, dense with glass towers. Most of the plans of the Internet entrepreneurs gathered in the Columbia Club tower depended on broadband communications. How on earth could all these vertical structures be served with fiber optics? he mused. Expensively, he thought. The estimate at the time was some $300 thousand per building, with tangled wires in conduits under the streets and along the walls and up the elevator shafts. Government auctions impended for LMDS microwaves, but Amadon wondered where the roof borne antennas could be placed and whether this microwave bandwidth, restricted to the low megabits per second, would suffice in the face of an explosion of Internet traffic. He suddenly arrived at the thought that the real next big thing would be optical wireless cellular communications through all those glass windows.

The others at the luncheon raised the usual objections: Fog, snow, rain, sleet, and turbulent air, even the tinted windows themselves would shrink the distance the signals could travel to an unusable few feet. The lasers would burn out people's eyes. Similar objections afflicted the early years of fiber optics, when experts concluded that impurities in the fiber would bar long-distance communications. But Amadon was not convinced. A 1973 Stanford graduate in political science, he had become infatuated with holograms as a student and had built himself a holographic laboratory to contrive those three dimensional images. Then he had gone on to become a White House correspondent during the Carter years and a cameraman for CBS News.

Foreshadowing his later epiphany, one of Amadon's jobs was to cover Ronald Reagan at his ranch outside Santa Barbara. The young amateur engineer set up a telescopic camera on a promontory some five miles away and took a famous picture of a waving Reagan that the President later signed: "There's nothing like being in the public eye." Today, TeraBeam uses both holographic and telescopic technology. Amadon invented a holographic lens to compress telescopes into a few score centimeters. These proprietary devices, manufacturable for around $150 apiece, could send and receive infrared beams and focus them on tiny photodetectors that could be coupled to fiber optic cores a few microns wide.

Amadon's invention was exactly on target for the telecosm and indeed he would find his first major funding for the company at the 1999 Telecosm conference.

The key to the telecosm is the shift of the focus of ascendant technology from the particle side to the wave side of the quantum duality, from the solid states of silicon to the spectronics of communications. The ruling realm of waves is the electromagnetic spectrum, which defines an essentially infinite span of frequencies, measured in hertz or cycles per second that begins with DC or direct current (no cycles at all) and runs all the way up through cosmic rays at the top measured in the petahertz (10 to the 15th). So far wireless communications has used well under one percent of the available frequencies.

For a decade, I have been awaiting the consummation of this spectronics paradigm. Higher frequencies bring shorter wavelengths, smaller more directional antennas, less interference, more capacity, and more bandwidth. More bandwidth, as Claude Shannon showed in his paper launching communications theory in 1948, requires less power. In the new paradigm of portable, handheld, fiber optic, battery or solar powered equipment, power would prove to be the most binding constraint and scarcity. Radio engineers have preferred a regime of long and strong, high powered large wavelengths that can bounce around the globe, scuttle along the ground, and penetrate walls. They have moved up spectrum only when there was no other way to find more capacity.

For example, the two-way wireless industry began at 100 megahertz and accommodated a few hundred ambulances and police cars. Moving up to 400 megahertz in the mid 1970s, the special mobile radio systems served tens of thousands of taxis, delivery vehicles, and other business communicators. In the 1980s, the industry ascended to 800 megahertz and scored millions of users around the world with analog cellular systems. In the 1990s, Personal Communications Services (PCS) climbed up to a band at two gigahertz and moved toward the hundreds of millions of subscribers. In the new millennium, the entire spectrum is opening up and the harvest is rapidly approaching the billions.

For nearly a decade, I have stressed the ultimate promise of optical frequencies in the air and even entertained the idea of ultraviolet. But experiments in wireless optics have been mostly restricted to the military and small enterprises in Israel. At the 1997 Telecosm Conference, Canon announced but refrained from demonstrating its optical Canonbeam. At the 1998 Telecosm, David Medved of Jolt in Tel Aviv explained the physics of over the air optics and described his company's several successes and customers. In 1999, Robert Martin, chief technical officer of Lucent, announced the Optic Air Wavestar product. In January of 2000, the Lucent device made a stir when it was used at the Superbowl to send uncompressed HDTV signals at 1.5 gigabits per second from a camera in a truck a kilometer away to the ABC hub at the Georgia Dome.

Neither Jolt nor Lucent, however, introduced a fully telecosmic cellular product. Optics cannot be fully disruptive until it can serve thousands of customers in a particular locality at a cost competitive with wireline systems, microwaves, fibers, and power lines. Several other companies offer interesting point-to-point, line-of-sight technologies that demonstrate the utility of these optical infrared frequencies. But all of these point-to-point networks are limited to special users, like governments or TV broadcasters. They are uncompetitive for most companies or consumers. For upspectrum companies serving ordinary business buildings, we have long had to settle for the 24 gigahertz, 28 gigahertz, and 38 gigahertz of Local Multipoint Digital Service (LMDS), all far below the 200 terahertz infrared region where fiber, and now fiberless, optics operates. With its obnoxious auctions, exclusive spectrum assignments, contentious and crowded rooftop real estate, and radios that would always be ready next year, LMDS was mostly a default paradigm. Finally, now, the LMDS episode is over. The TeraBeam era begins.

To achieve this breakthrough, TeraBeam had to invent both a cellular hub and a customer device that could be produced affordably for scores of thousands of customers. In the process, the team led by Amadon and his colleague Richard Rallison concocted some seven major innovations.

Crucial was a point to multipoint downstream technology from the base station or hub to the customer premises. Without point multipoint, all the expensive optics would have to be duplicated for every user at both ends—more or less creating Lucent's Superbowl setup for every connection. At present, this is what most LMDS deployments do. They use a pair of point to point radios for every customer—a transceiver in the hub and a transceiver on the roof of the office, like a set of CB radios or walkie talkies. If wireless optics used that topology, it would be limited to expensive applications for elite customers.

In a compact and cost-effective chassis, the TeraBeam hub at full capacity broadcasts encrypted signals downstream to all users, while keeping track of point-to-point upstream links from each customer site separately. Each downstream floodlight can serve one 90 degree cell sector which contains potentially scores of customers. The customers identify their own messages by an IP (Internet Protocol) address, then decrypt them in real time, and send responses back to be detected and routed by the hub.

As in WDM fiber optics, the most crucial component is the Erbium Doped Fiber Amplifier that can enhance hundreds of separate wavelengths or bitstreams at a time. In fiber optics, the EDFA is installed in the fiber itself, to enhance the signal and extend the distance it can travel without being electronically regenerated. In the TeraBeam system, the basestation hub houses the EDFA. There it takes the low milliwatt signals from the network and ratchets up their power to a level where they can be broadcast through a cell with a diameter of three kilometers or more.

Without this amplifier, it would be necessary to regenerate each signal separately, duplicating this costly and complex device for every customer. For each gigabit link, the amplifier must provide a 200 milliwatt signal (for lower bandwidth connections, the 200 milliwatt signal can be split into signals of 20 milliwatts or less). Some 200 milliwatts for each of 24 users in a sector adds up to five watts. TeraBeam's problem was apparently a showstopper. Oriented toward the milliwatt powers of fiber optics, the industry did not manufacture five watt EDFAs.

High power EDFAs depend on the external pump laser that powers the device. The key manufacturers of pump lasers are JDS Uniphase and SDL, but even the strongest pump lasers from SDL operate at close to two watts. SDL's multiwatt pumps are now in demand for undersea applications and for Raman Amplifiers needed in the long distance fiber systems now being rolled out by Corvis and Q-tera (Nortel) to reach 3,000 kilometers without regeneration. But SDL's pumps do not suffice to propel Terabeam's signals three kilometers through the air.

On the Terabeam advisory board, however, is Valentin Gapontsev, a burly saturnine expert in optics, who in a previous career in Russia specialized in high powered lasers used to cut sheet steel or intercept missiles. Now CEO of IPG Photonics in Sturbridge, Massachusetts, he runs the world's only company that can make pump lasers that operate at 5 watts. With high powered uses multiplying in communications, IPG should become increasingly important to the Telecosm. TeraBeam has negotiated an exclusive contract with the company for high powered EDFA modules used in broadband wireless optics. Problem solved.

The rest of the TeraBeam hub uses lasers and photodetectors similar to those found in fiber optics systems. Focusing on Internet uses, TeraBeam's routers and switches employ gigabit ethernet to transmit IP packets. Using advanced telescope optics and unique high powered amplifiers in the transmitter, Amadon's strategy is to overdesign the hubs in order to enable cheap and robust systems at the customer premises.

It is at the customer premises that the full novelty of TeraBeam becomes evident. Crucial is a compact telescopic antenna design that allows transmission through windows without endangering the eyes of anyone who happens to look into the beam. This requirement meant giving up the previous idea of using cheap 780 nanometer lasers manufactured by the billions for CD players. The 780 wavelength is so close to the frequency of visible light that it readily enters and damages the human retina. The 1550 nanometer wavelength used in fiber optics, however, is too large to damage the eye at the power levels used by TeraBeam. At 1550 nanometers, moreover, the tints commonly used in office building glass do not block the signal.

Also needed was a large collecting aperture to receive the light, yet not too large to be readily placed in a building window. Made of a proprietary material, Amadon's holographic optical element costs $150 to build and eliminates all the space hungry and expensive curved lenses that raise the cost of a comparable telescope to $12 thousand. It enables the 16 inch collector to be incorporated into a receiver no larger than a DirecTV antenna.

Enhancing the system is a pointable and power controlled beam shaping technology that allows the transmission of narrow signals to the most demanding customers or into sectors with too few users to justify "flood lighting" the entire space. The directionality of this kind of light permits unlimited frequency reuse since lightwaves can be located side by side within the 16 inch span of the 96 segmented antenna. By contrast an LMDS signal would need a 150 foot diameter antenna and elaborate insulation to prevent the different signals from interfering with one another.

The light is narrowly focused to a diameter of 120 centimeters. Sophisticated tracking technology enables adjustment to moving targets such as skyscaper towers shifting in the wind.

TeraBeam takes the advances of WDM optics and moves them to the local loop. At the outset, WDM will be restricted to the company's backhaul systems linking to the larger fiber networks. Relieving any need to use many lambdas in the cell itself is TeraBeam's ability to reuse the same frequency multiple times through juxtaposing the narrow beams of light without interference. But in the future, TeraBeam's technology can also use WDM to multiply capacity and flexibility with each customer.

The company has a $450 million investment from Lucent. Initially TeraBean's chief competitor—with Optic Air—Lucent investigated Amadon's technology and resolved to switch rather than fight. Lucent will own 30 percent of TeraBeam's manufacturing arm and will sell gear to TeraBeam proper on concessionary terms. TeraBeam's business plan envisages a cost per sector for a one gigabit–per-second initial system as $26,000 a month, dropping to $12,000 over time, including the customer's equipment. Anticipated monthly revenues per customer are $6,300, hence break even comes at two customers per sector. The cost per bit per month is roughly one thousand times less than a T-1 line and installation takes minutes rather than months. A full load of 24 customers per sector will bring in $150,000 per month. Using splitters, WDM, and other enhancements lowers costs, raises capacity and multiplies potential revenue. A scalable system, TeraBeam will be able to expand both its bandwidth and capacity in accord with demand.

Planning to become a major player in US cities, TeraBeam hired as CEO Dan Hessey, a 25 year veteran of AT&T who ran its Wireless Group, originally McCaw Cellular, then about to be spun off in an IPO that would have yielded him a possible $20 million in options. Instead, Hessey will lead TeraBeam into a direct competition with AT&T and all the other companies competing to serve businesses over the last mile. TeraBeam initially commanded eight hubs in Seattle and is rolling out service in fifty cities through 2002. It also contemplates alliances with network partners such as Level 3 and Global Crossing.

With wireless optics, the Telecosm is consummated from the backbone to the business market. Wireless surprises over the next decade will sweep into homes and small offices as well. The world economy will ride the light as much in the air as in fiber.

TELECOSMTELECOSMTELECOSMTELECOSMTELECOSMTELECOSMTELECOSMTELE-

COSMTELECOSMTELECOSMTELECOSMTELECOSMTELECOSMTELECOSMTELE-

COSMTELECOSMTELECOSMTELECOSMTELECOSMTELECOSMTELECOSMTELECOS

MTELECOSMTELECOSMTELECOSMTELECOSMTELECOSMTELECOSMTELECOSMT-

ELECOSMTELECOSMTELECOSMTELECOSMTELECOSMTELECOSMTELECOSMTELE-

COSMTELECOSMTELECOSMTELECOSMTELECOSMTELECOSMTELECOSMTELECOS

PART FIVE

MTELECOSMTELECOSMTELECOSMTELECOSMTELECOSMTELECOSMTELECOSMT-

ELECOSMTELECOSMTELECOSMTELECOSMTELECOSMTELECOSMTELECOSMTELE-

THE MEANING OF THE LIGHT

COSMTELECOSMTELECOSMTELECOSMTELECOSMTELECOSMTELECOSM

TELECOSMTELECOSMTELECOSMTELECOSMTELECOSMTELECOSMTELECOSMTELE-

COSMTELECOSMTELECOSMTELECOSMTELECOSMTELECOSMTELECOSMTELE-

COSMTELECOSMTELECOSMTELECOSMTELECOSMTELECOSMTELECOSMTELECOS

MTELECOSMTELECOSMTELECOSMTELECOSMTELECOSMTELECOSMTELECOSMT-

ELECOSMTELECOSMTELECOSMTELECOSMTELECOSMTELECOSMTELECOSMTELE-

COSMTELECOSMTELECOSMTELECOSMTELECOSMTELECOSMTELECOSMTELECOS

MTELECOSMTELECOSMTELECOSMTELECOSMTELECOSMTELECOSMTELECOSMT-

ELECOSMTELECOSMTELECOSMTELECOSMTELECOSMTELECOSMTELECOSMTELE-

COSMTELECOSMTELECOSMTELECOSMTELECOSMTELECOSMTELECOSM

TELECOSMTELECOSMTELECOSMTELECOSMTELECOSMTELECOSMTELECOSMTELE-

COSMTELECOSMTELECOSMTELECOSMTELECOSMTELECOSMTELECOSMTELE-

COSMTELECOSMTELECOSMTELECOSMTELECOSMTELECOSMTELECOSMTELECOS

MTELECOSMTELECOSMTELECOSMTELECOSMTELECOSMTELECOSMTELECOSMT-

Chapter 18

The Lifespan Limit

As free economies banish the age-old pressures of material shortage, as rapid transport and communications dissolve the bounds of geography, as capitalism increasingly crowns every producer, every engineer, every worker, every customer with the sovereignty of abundance, the ingrate heirs of all this plenty look up at last and contemplate the residual scarcity: the tyranny of the clocks, the limits of the span of life.

Time's winged chariot, of course, has long been the stuff of poems. But as the prime fact and canonical scarcity of today's economy, it is an absolutely new concept. During the long millennia of material scarcity, the customer's time was what economists call an externality, like air or water. It was an economic asset so readily available that it escaped economic accounting. In the old economy and a holdover in the new, a key rule of commerce was: Waste the customer's time. This was not an accident or a mistake. It was so close to the heart of marketing strategy that people long took it for granted, like slavery in the precapitalist era. They hardly even noticed how prodigally businesses wasted their time.

In the era of the telecosm, people will shake their heads in amazement that we ever put up with this time-wasting regime. The new rule is: The customer is sovereign and he knows what he wants: It is not your product; it is time.

Most of today's businesses still follow the old rules. Of course, we waste your time. Make you drive to the bank, grocer, bookstore, post office, hospital, telephone company, realtor, broker, library, software vendor, city hall, school. Make you line up in a queue. Fill out forms in triplicate. Read legal documents, written in unfathomable jargon, and sign them mendaciously in seven places, in order to buy a house or a company. Make you clip coupons from newspapers to qualify for discounts. Lure you into the state lottery and make you hang on tenterhooks awaiting the results. Make you enter a gauntlet of prescripted charades with a congeries of slack and smarmy pettifoggers in sales, insurance, finance, and government in order to purchase a car. Drag you from the tub to answer a telemarketing cold call from someone who mispronounces your name and thinks you want to buy municipal bonds right away, or a subscription to *Urology World,* or make a contribution to Harvard, or purchase a special discount long-distance card from Tabitha's Beauty Salon and Telecom.

Your children, meanwhile, go to a school or college which focuses not on teaching them what they need to know to prosper in the world but what a set of bureaucrats think they should know to adapt to the ideologies of the media and academy. In California, for example, some 40 percent of all school funds go for fifty-one categories of education excluding the three Rs, history, and sciences, but including special ed, environmental ed, bilingual ed, sex ed, teen pregnancy ed, drug ed, cultural awareness, and gang intervention.

Prohibited from working—that is, learning on the job—students sit in classes with a score of other students who each have different aptitudes and educational needs. They must perforce adapt to the pace of the slowest. They fill out forms and stand in queues and squander time. They practice putting a condom on a banana. In the end, they will acquire most of their education later in life at their place of employment just as you have. But the U.S. educational system, comprising 15 percent of GDP and costing perhaps a trillion dollars, is maladapted to supplying the lifelong learning demanded by the new economy. It wastes the student's time, just as the government wastes the citizen's time and businesses waste the customers' time.

For business and government, your time is an externality, and you get used to it. They line you up and bore you to tears. You take it, shifting from one foot to another, thumbing through the newspaper, until you get to the front of the line, where they tell you they can't help you until you fill out the correct documents and join the queue in the office across town. Even the telephone is a huge time waster. Not only does the average customer around the world wait 38 months for the installation of a line, but it is a hit or miss medium. Half the time it does not deliver the wanted party when you call and half the time it inflicts an unwanted intruder when someone calls you. Telephone tag, phone mail jail, dropped connections, telemarketing invasions are all regarded as an acceptable cost of phone ownership.

The supreme time waster, though, is television. Many people still have trouble understanding how egregious a time consumer, how obsolete a business model, how atavistic a technology, and how debauched a cultural force it is. You sit down on a couch in front of a screen, to watch degrading and titillating lowest-common-denominator trivia, scheduled for you in some netherworld between Madison Avenue, the FCC, and Hollywood, offering a sordid stream of sleazy banalities, gun grunge, bedroom mayhem, and offal innuendoes, some preening as "news" and some leering as entertainment, for as much as seven hours a day, on average, consuming perhaps two thirds of your disposable time, year after year, all in order to grab your eyeballs for a few minutes of artfully crafted advertising images that you don't want to see, of products that you will never buy. Is it a breast? Is it a thigh? No, it is a fender! A frosted Beemer? No, a beer bottle. TV ads that are as irrelevant to you, 90 percent of the time, as the worst telemarketing spiel. Justifying this scheme is the "free public service" that television supposedly offers, namely the "serious" portions of the "news" (chiefly government propaganda) and Saturday morning children's programming (more propaganda).

And you take it all in stride, because wasting your time has been until recently the essence of business. You spend two hours installing programs on your personal computer and then another half an hour waiting on the phone for technical support because the program still issues enigmatic error messages, until on reaching the number at last, you enter a maze of service options that seek to make you waste still more time solving the problem yourself by poring through a large body of usually irrelevant material, either in a cumbersome manual written in Japanglish or on an only slightly more coherent web page. Finally you make your way to the voice of a young man on the other end of the line who cannot hide his disdain for you or doesn't really try, asking whether you have mailed in your customer registration card, and what, among the array of gnomic digits imprinted on the documentation is your customer number? Finally, at least a third of the time, if he deigns to address your problem, he tells you it is the fault of some other vendor.

I do not even want to tell you what happens when you call your phone company for installation of a broadband link to the Internet. Or rehearse all the costly and endlessly time-consuming indignities—adding up to several billion hours annually—inflicted on U.S. citizens in the name of collecting taxes nearly as inefficiently and intrusively as possible. In all, U.S. citizens spend some 7 billion hours every year filling out government forms. The fact is that the entire economy is riddled with time-wasting routines and regimes. Suffice it to say that the concept of the customer's lifespan as a crucially scarce resource, indeed the most precious resource of the information economy, has not penetrated to many of the major business and governmental institutions in the U.S., let alone overseas.

The message of the telecosm is that this era is over. Its corpse will remain around the house for another decade, emitting unsavory fumes, and millions of people will continue to regard it as a source of wealth and happiness, but it's as dead as slavery in 1865. Just as air and water entered accounting ledgers and public consciousness only when the abundance of goods and wastes of the industrial age came to be deemed pollution, time has become the customer's most precious resource as material abundance is taken for granted. The customer who is well fed and sheltered, the customer who senses the possibilities of the new technologies, is no longer going to put up with standing in unnecessary lines, filling out gratuitous forms, telling telemarketers whether he has had a nice day, and waiting for bureaucrats to get around to his case. In an age of affluence, lifespan—that is time—becomes the ultimate scarcity against which the value of government policies, companies, and commodities are measured.

The first sign of the change is the ongoing increase in concern for health. The extension of the average lifespans in the industrial world by some 10 percent over the last forty years is a crucial index of economic advance, while the simultaneous decline of average lifespans in the Soviet Union revealed the utter emptiness of Communist claims of economic success. Furthermore, the share of national resources devoted to health care is rising in all industrialized countries. Although the increase partly reflects poorly designed medical insurance schemes, this shift is a signal of the new priorities of affluence. Unless excessively regulated by government, creating time-wasting monopolies, health care companies will be prime winners in the new era. Huge opportunities lie open for health care providers that develop new ways to deliver services at a lower cost in the user's time.

The microcosm chiefly redistributed power within existing companies and facilitated the formation of new companies. It gave an individual at a workstation the creative power of a factory tycoon of the industrial age. Corporate CEOs who once could perform most of the jobs in their companies now command only a small portion of their firm's base of knowledge and skill. The telecosm goes further, and distributes power out of companies to their customers. It gives every customer the ability to survey the entire global marketplace and make optimal purchases—a power over the market that in the previous age eluded even the most sophisticated procurement offices of large companies. Together the microcosm and telecosm give a single individual hooked up to the Internet the power not only of a factory tycoon of the industrial era but of a broadcast magnate of the television era.

The Internet saves the customer's time. Because many experts continue to see the customer's time as an externality, outside of productivity data, they miss the profound impact of this technology. They will not have a clue about what's going on until the revolution is already almost over.

Consider the customer of the new era. He awakes in the morning to the music of his teleputer playing his favorite wake-up matinade through a speaker near his bed. On a screen across the room is the news he needs. One of his stocks has lost three points. He clicks and a full report emerges. It is written in a font with resolution as good as paper. He doesn't have time to read so he clicks on audio. While he shaves, a voice intones a report on the news affecting the stock. Then he clicks again, transferring the report to his broker with an inquiry about appropriate action. Or perhaps he does not have a broker and he decides to average up on the shares by taking advantage of the lower price to buy more. He can accomplish these tasks with another few spoken instructions, avoiding both phone tag and search time. They happen almost spontaneously as he goes through his morning routine.

Meanwhile his wife consults another screen for the local weather and school reports to find out if the snowstorm during the night has caused cancellation of classes at the high school one of their teenagers attends. The younger two children are being taught at home through the K–12 teleputer curriculum that links them to courses across the country that are suitable for their age and aptitudes.

Their son enters the room complaining of serious stomach pains. He is referred to an Arthur Robinson diagnostic sensor linked to the Internet. He breathes into the tube (or supplies urine to it) and by a rapid reference to a massive database of health correlations, his condition is appraised and the results dispatched to a doctor. Within minutes, the doctor responds with an assurance that the illness is not grave and prescribes an appropriate medicine for the boy. The doctor is paid on the spot through the web.

The husband sees on his personal news summary that his college track team won a close meet the night before. He clicks on a highlight video which shows the thrilling final quarter of the mile relay race; a reasonable extra payment is deducted from his account. In the era of the telecosm, professional sports events will still make their claims, and the husband may well continue spending time with his favorite teams as he did when television ruled. The Super Bowl is coming up, after all. But these mass events may decline in relative significance, compared to private games and competitions. His high school daughter, for example, shuns the WBA, prefering to watch her older sister perform in a college cross-country ski race. She clicks to a recording of this event on yet another screen and takes in a few minutes as she eats breakfast. Then she is off to school with a printout of her calculus teacher's corrections to the problems she had submitted two days before over the Internet.

The man clicks to his finance program. He wonders how much taxes he has paid to date under the new flat tax. Self-employed, he receives most of his income by encrypted electronic transactions—allowed under another new law. Without any way to measure incomes in an era of encrypted money, the program automatically makes the transfers of funds based on purchases of goods and services. He still thinks taxes are too high and resolves to invest more of his money, since investments escape the burden of a tax focused on value added or consumption.

Meanwhile, by 9:00 A.M., the wife has done all the grocery shopping for the day. By a long-distance teleconference, she and her husband visited her elderly parents in Hawaii and took a virtual walk with them on the beach. They have read all the news they wanted—with videos included where appropriate—and are ready to begin their jobs in the next room. In one hour in the morning, they have accomplished tasks that would have taken most of the day under the previous regime when span of life was still an externality.

As a result, they have more free time. They might spend the extra time socializing, or doing more work. Many pundits will express the usual alarm that people are working harder than ever. This is a "problem" that the Internet will not solve. Since human beings through history have thrived through work, most people will use their liberated time to perform more valuable economic activity. Using the web, they will be able to work far more efficiently, collaborating with the top experts everywhere and serving markets around the globe. "Cottage" industries will burgeon and many of them will become important corporations. Liberated from hierarchies that often waste their time and talents, people will be able to discover their most productive roles and pursue them more effectively. Under capitalism, where profit comes from serving others, this release of entrepreneurial energy will be more morally edifying than the "leisure" diversions that many imagine to be the end and meaning of life.

In the evening, the family may decide to watch a film together. Choosing from hundreds of thousands of possibilities, sorted and ranked by content and appraised by reviewers they trust, they decide to "attend" the global opening of a new movie. A few vocal clicks on the teleputer and it is immediately served up on the World Wide Web. Video will be nearly as diverse and as rich in cultural content as book culture is on today's Internet, with millions of titles available at your fingertips at Amazon or Barnes & Noble or their hundreds of more specialized affiliates.

In all this activity of the future, advertising will remain as important as ever. After all, there must be ways of publicizing new products in an ever more innovative economy. But you no longer will have to look at ads that you do not want to see. In the new economy, this change will be most difficult for advertisers and marketers. Madison Avenue has already been flailing about, contemplating ways of paying people to watch their ads, with coupons, advantage miles, movie tickets, and lottery slips. Thus they admit that these exhibits are not ads at all; they are minuses.

In the future, no one will be able to tease or trick you into watching an ad. Your time is too precious and you are too powerful. Advertisements will truly add value rather than subtract it. You will look at them when you want to buy some thing or as an accessory to your personalized news. Because the ads will relate to your interests, they interest you. They come not as intrusions but as attractions. If the product intrigues you, you can seek further information on it. In most cases, you will be able to move step by step toward an electronic purchase, summoning information as you need it from a variety of objective sources.

Today's television watchers resort to zappers and remotes and other devices to avoid advertisements as much as possible. With cable, you actually pay extra to receive ad-free programming. But readers of computer magazines actually seek out the ads. Indeed, most of their subscribers would rather pay extra for issues with ads rather than extra for those without. The reason for the superior effectiveness of magazine and web advertisements is that they are targeted. The consumers of these publications have a community of interest. They are already owners of Macs or PCs and they are interested in enhancing their property.

With targeted advertisements, marketers assume substantial knowledge on the part of their viewers. Disciplined by their customers, they must be objectively informative. The ultimate model is the infomercial—the commercial program that is eagerly watched by the viewers. Because TV infomercials must address a broad lowest common denominator audience, they currently have a down market orientation. But the concept itself is not down market as the comparison to technical publications shows.

In a world where advertisements are read only by people who choose to read them, the quality sharply improves. Freed from the meretricious need to appeal to passing strangers, advertisers ascend to a higher cultural level. The hokiness and smarm we associate with Madison Avenue dissipates and the advertiser makes an honest effort to convey as much information as possible in as compelling as possible a style. Advertising becomes an art of truth rather than an art of deceit.

As the World Wide Web moves beyond the broadcast advertising model—the billboard flash at the top of the screen—advertisers will have to stop wasting the customer's time with teases. When they do, they will discover that targeted ads are hugely more effective and can support more programming, including free services. Today television purports to be free, but in fact we pay through the nose for these gratuitous distractions, in wasted time, cultural decay, educational decline, family breakdown, criminal activity, and aesthetic degradation. The manifestation of real freedom is choice.

The greater efficiency of targeted advertising springs not only from customers' superior knowledge of products but also from advertisers' superior knowledge of their customers. As Don Pepper and Martha Rogers observe, by using database tools on the web a company can now remember each customer as readily as a customer can remember a company.

Many observers fear this increase in advertisers' knowledge as an invasion of privacy. This view misconceives the nature of privacy. Invasions of privacy reflect inadequate knowledge, not excessive knowledge. In a sense, telemarketers and other advertisers fail to invade our privacy enough. Someone calling you at the dinner hour to sell you a product of no interest is intruding on your time out of ignorance. This ignorance may even increase as a result of laws that make it more difficult for companies to collect information about their customers, and what they want, and how they prefer to be alerted to new products. Most customers are pleased to be surprised by advertisements of products that they want.

In general, the threat to your privacy and the menace of intrusion on your life does not come from true information. Throughout much of history people have lived in small towns and villages where your reputation could be destroyed without recourse because someone imagined that you were a witch. By leaving a trail of encrypted information about purchases and activities, even if much of the data is captured by businesses that want to sell them goods and services, people can maintain control of the truth of their lives.

Needless to say that the Internet will not solve all the problems of human life. People will continue to suffer and die, continue to sin and pay, continue to waste time and wealth. But businesses will no longer see their customer's time as an externality; it will be the central test of commercial viability. Does this new good or service reduce the wavelengths and frequencies of tedium and trivia and pettifoggery in the lives of their customers? In pursuing the new requirements of the telecosm, businesses will adopt new means of marketing. They will adapt their businesses to the new king and sovereign: the customer.

Chapter 19

The Point of Light

Imagine gazing at the web from far in space. To you, through your spectroscope, mapping the mazes of electromagnetism in its path, the web appears as a global efflorescence, a resonant sphere of light. It is the physical expression of the converging telecosm, the radiant chrysalis from which will spring a new global economy.

The luminous ball reflects Maxwell's rainbow, with each arc of light bearing a signatory wavelength. As the mass of the traffic flows through fiber optic trunks, the skeleton glows in infrared, with backbones looming as focused beams of 1,550 nanometer radiance running across continents and under the seas. As people more and more use wireless means to access the net, this infrared skeleton grows a penumbra of microwaves, suffused with billions of moving sparks from multimegahertz teleputers or digital cellular phones. Piercing through these microwave penumbra are rich spikes of radio frequencies confined in the coaxial cables circling through neighborhoods and hooking to each household. Spangling the penumbra of the net are more than a hundred million nodes of concentrated standing waves, each an Internet host, a computer with a microprocessor running at a microwave frequency, from the hundreds of megahertz to the gigahertz.

The radiance reaches upward between four hundred and eight hundred miles to thousands of low earth orbit satellites, each sending forth cords of "light" between earth and sky in the Ku band between 12 and

18 gigahertz. Interlinking the birds of Craig McCaw's and Bill Gates's Teledesic will be a halo of 60 gigahertz radiations.

Then imagine that every hundred days the total brightness doubles. Not only does the total number of screens rise by more than one third but also the traffic on each of the links rises by 50 percent. Federal Express ignites a flare in Memphis and its glow swells into the millennium. The two thousand hotels of Marriott go on the web for the first time. That pulse of light flashed in 1999. AOL customers leave behind their 28-kilobit links and move to 56 kilobit and ethernet modems. A larger surge ripples across the ball of light. Corporate ethernets leap up from 10 megabits a second to 100 megabits and then to a gigabit and 10 gigabits, and exfoliate intranets and extranets in ardent loops. RealNetworks moves to full-screen video streams in a global piñata of luminous threads. Cable modems raise tenfold the average bandwidth of network connections to homes and schools. Digital subscriber loops add to the fuzzy glow on the edge. Microsoft's NetShow spurs a surge of video teleconferencing. Digital cellphones begin sending and receiving video images and storing voices. All pump up the lumens of the encircling radiance.

As the intensity of the light rises—as more and more photons of traffic flash through the webs of glass and air—the change pushes up the overall frequency or average color of the light. Traffic on cable coax runs at radio frequencies. When it moves into glowing silica threads, it leaps upward in the spectrum. The lasers of fiber optics emit waves that cycle a million times faster than do the oscillators in a cable TV system. So as the brightness increases, its average color also inches up the spectrum. The global iridescence changes its dominant hues. If it were a rainbow, the center of intensity would move up from red through green toward violet. If it were a meteor, the Doppler blue shift of the Internet would suggest it is approaching you.

The change in "colors" signals a shift in the nature of the network. Rising toward the terahertz region, colors become more and more difficult to manage electronically. So as the core of the net becomes brighter, it also necessarily becomes dumber. Signifying the declining electronic intelligence of the net is the sharp drop in the number of relatively low-frequency devices scattered among the webs of infrared optics. The electronic switches and routers and cross connects give way to passive optical shunts and filters. Electronics gives way to photonics.

As the networks become dumber, though, they attract more and more smart entrepreneurs on their edges. Businesses are freed to focus on ingenious products rather than on clever networks, free to focus on content software rather than on conduit protocols, free to explore the intricacy of more effective search engines and data retrieval systems rather than the complexities of adapting voice circuits to mazes of data types.

Thus the telecosm inscribes its signature on the space station spectroscope: the growing brightness and the rising colors mark the declining intelligence in the core of the web. The spectrogram of the Internet reveals the evolution of the network economy.

Through these coruscating channels will ultimately run most of the commerce of the world; more value will move by resonant light than by all the world's supertankers, pipelines, eighteen-wheeler trucks, and C5A airships put together. Yet all these frequencies, visible on the spectroscope in space, are invisible to you and me on earth. The Internet is a cloak of many colors for the communications of the world. But human senses can grasp none of its tints and spangles. The 400 terahertz of visible light are absent in the links. On the global ball of light of Internet frequencies, visible light sparkles only in the billion phosphorescent spots where there is a screen.

Shining on those screens are the images of a new economy: the cornucopian spiel of an eBay auction, the shuffling of cards and companies in a Yahoo portfolio, the spray of Fourier transforms in a DSP class, the specs and span of National Semiconductor's web-based inventory, the targeted immediacy of a streaming web cast, the one-click ignition of an Amazon transaction, a buxom blooming of Laetitia Casta wallpaper, the tintinnabulation of cash registers at Cisco and Dell, the quick click of an E-mail attachment, the flaunt and vaunt at a virtual bar in a teen chat center, the rustle of Hilton reservations, the theatrical growl of Tennyson's *Ulysses,* the populous skies of a Java airline map, the sepulcral murmur of a Leonard Cohen meditation, the mad snorting and pawing on a Roaring Bull thread, the swashbuckling plays of a multicity game. (And, yes, a stygian undercurrent of oily nude body parts for the sticky keyboard set.) Creeping slowly across the radiance, and enhancing it, are the harbingers of full-screen films, three-dimensional images, and high-resolution teleconferences. When they become common, the luminosity will surge another millionfold.

Seeking clues to the meaning of this radiant transformation of the globe, I began by thrashing through haystacks of economic literature. I was looking for a needle of insight through which to march all the gaudy caravans of undulating camels, laden with Internet glow. . . . or was it Internet hype, bumping and grinding for a manic Wall Street? In the end, I found a pin instead.

The pin was a unit of output of the factory in Adam Smith's *Wealth of Nations,* exemplifying the power of mass production and the division of labor. Smith showed that a worker specializing in one part of the process of manufacturing pins could be hugely more productive than a worker attempting to produce pins entirely by himself.

Living in a time of transition resembling today's, Smith was boldly explaining the contested passage from the age of guilds and crafts to the age of mass manufacturing. Smith showed that the factory workers of the industrial age were not two times or ten times more productive than craftsmen were, but ten thousand times more productive. To do an accurate valuation of the economy at that time, you would have had to apply this ten thousandfold factor to the sectors of industry that were adopting the new technologies. Instead, while indices of wealth and longevity surged upward, many sensitive souls focused on the clutter in the streets and the smoke in the air as millions thronged the cities to take the new jobs in what Blake called "dark satanic mills."

A key reason that specialized workers can each produce ten thousand times more than the single craftsman is their faster process of learning. Each worker has to master only one part of the process. Because he gets to apply his intelligence to that one component more frequently, he accelerates his learning.

In the time of Adam Smith, the workers could not gain the ten thousand fold edge without coming together in a single factory at a single time. Capital and equipment was scarce and costly. Imagine, though, a global cornucopia of capital flowing readily across borders and through the ball of radiance. Imagine a cost of equipment, based on microchips and fiber threads, plummeting to less than a dollar for a million instructions per second or $19.95 for thirty days of access to Internet exabytes. Imagine that any worker could collaborate with any other worker at any time—that the factor of ten to the fourth could be harvested by any collation of workers in a virtual web anywhere on earth. That is one measure of the meaning of the light. The nodes of creative effort could summon their ten thousandfold magic of learning at will in minutes rather than in years.

An incandescence rising up with little overall planning or guidance, the Internet raises again the riddle faced by every sophomore physics student: the rise of civilization in the face of the law of rising entropy. The second law of thermodynamics, the entropy law defines the tendency of all order and energy to deteriorate into disorder and waste. Is the ball of radiance a meteor that will burn out, leaving only darkness and ashes? So seems to say the second law of thermodynamics: entropy increases. Things get worse. Energy is used up and transformed inexorably into lukewarm gases.

More relevant to the Internet, Claude Shannon used the same word—*entropy*—to designate information content in a communications channel. More entropy in Shannon's code signifies more information. In Shannon's terms, entropy is a measure of unexpected bits, the only part of a message that actually bears information. Otherwise the signal is only telling you what you already know.

To send unexpected bits—a high entropy message—you need a low entropy carrier: a predictable vessel for your meaning. You need a blank sheet of paper that does not alter or obscure the message inscribed on it. For a microchip, crystalline silicon offered the regularity and purity that was needed to bear the intricate designs of an integrated circuit. For long-distance communications, the perfect carrier is an electromagnetic sine wave: a perfectly predictable sideways S curve. By distorting the wave in a systematic way—by modulating it—you can transmit information. By removing the predictable part of the signal, the original sine wave, you can capture the information, the unpredictable part, at the other end.

In order for the message to be high entropy (full of information), the carrier must be low entropy (empty of information). In the ideal system, the complexity is in the message rather than in the medium. By eliminating the entropy from networks—moving them onto the pure waves of light—you can increase their ability to bear information. Another word for a low entropy carrier is a dumb network. The dumber the network the more intelligence it can carry.

With more dumb carriers, bearing ever more information, with ever less friction and resistance, the ball shines ever more intensely with cumulative loads of luminous learning.

Through this learning, civilization defies the thermodynamic laws of decline and fall. Moving into the abundance of light, the Internet is yet another demonstration of the triumph of intelligence over time and chaos.

Henderson's Boston Consulting Group demonstrated the learning effect for thousands of different items, from chickens, eggs, and tires to insurance policies and telephone calls; Bain and Company refined and extended it as the experience curve; Moore's law captures its effects in the production of microchips; Michael Rothchild presented evidence that it is a biological truth, applicable to everything from ants and bees to Amazonian slime molds; Raymond Kurzweil deems it a law of time and chaos by which time—measured by the incidence of salient events—accelerates in every evolutionary process (and slows down in every chaotic process, governed by physical entropy). Now time seems to be racing through the global radiance.

Imagine the mesh of lights—the radiance of sine waves—as an efflorescence of learning curves as people around the world launch projects and experiments without requiring the physical plant and equipment and the regimented workers in Adam Smith's factory. Without the overhead and entropy, noise and geographical friction, entrepreneurial creativity takes off. The network reduces all the proliferating protocols and intricacies to one, the Internet Protocol. Increasingly IP packets run directly on light running down glass.

Becoming ever more predictable, the net can carry ever more voluminous and high information traffic. Swept up into an ever more photonic radiance, the traffic can soar a thousandfold every three to five years, more than a millionfold in a decade. A miracle for our time dwarfing Adam Smith's pins, it means that any Internet company currently handles only one tenth of one percent of its potential volume in five years. If it keeps its share of the market, it must expect a thousand times greater traffic in half a decade. This cannot happen on ordinary networks. As the traffic overflows the smart silicon passages of conventional wires and switches—where it is compressed and coded and celled and queued and framed and corrected electronically—it must soar up spectrum. On the wideband boulevards of photonic glass, everything travels in IP on lambdas and the new companies of the net eclipse the old companies of the microcosm.

This is the tragic rhythm of capitalism, the fatal triumph of enterprise, the vicious circle of learning. But it is also the guarantor of freedom. The radiance does not migrate to one place and stay there, whether Redmond,

Washington, or Santa Clara, California. It reaches its point of resonance and then gives way to new flares of learning and creativity. Without a high entropy medium of noise forming barriers in the middle, the distinctions and categories of conventional commerce break down. Buyers, sellers, producers, consumers, financiers, factors, insurers, and savers merge across the radiant ball and both the words and the roles lose their edges.

I log onto Amazon or Barnes & Noble and a "cookie" on my computer informs the system of my presence. I play a game of literary preferences. It is a service of Amazon. But it allows the company to alert me to new products that I might like. If they are not just right, I correct them. From my purchases and preferences, Amazon learns how to serve other similar customers. I review a book that I have read. Amazon prints the review. From my purchases and the purchases of others, it contrives a best seller list that is updated every hour. It informs me of the exact performance of all my books. I sell my books from my own web page, collecting a profit, and Amazon fulfills the orders. I am an Amazon customer, supplier, investor, client, author, audience.

The buyers of web services supply their own capital in the form of their linked computers. They provide infinite shelf space on their disks. They give or sell their names, their time, their knowledge. They learn how to improve the product. The sellers of web services use the equipment of their customers and gain knowledge and time from the buyers. Each population learns from the other, is the other. The customers are the product and the product is the customer and both serve one another, in a rhythm of creativity between producers and users, a resonance of buyers and sellers in which the buyers also sell and the sellers also buy in widening webs of commerce. The resonance is the wealth and the light and there is no impedance in the middle. A single attractive invention that resonates with the needs of others can unleash a cascade of creativity.

Governing the light is this law of resonance. Supply creates its own demand, and the supplies and demands in themselves create a market. That is the source of Metcalfe's law. The exponential value of networks is not in the links; it is in the light at the end of the links, suffusing the edges of the web. The Internet is not merely a radiance of connections; it is a mesh of constant invention. It is not a web of ever proliferating desires or demands; desires can well proliferate without it. The net is a seine of collaborative production. It is the capitalist means of production and it has fallen into the hands of a billion learners and value creators.

Through endless essays of trial and error, capitalism teaches every venturer the rules of resonance, the laws of right and light. It reveals what efforts reverberate in the minds and hearts and hands of other producers—what ventures enlighten and enrich them. It ruthlessly filters out the ego trips and feckless tries and self-indulgences and investments of disguised consumption, and products that exploit and diminish their customers. The radiance of the net rides on trust: predictable waves of unsurprise that can carry the entropic ideas of enterprise. The net promotes the ideas that yield more than they cost—the ones that are worth more to others than to the producer, the ones that teach more than they lose. Only products that enrich their customers ultimately create the kind of market that compounds its gains.

On the map of incandescence, light is not evenly distributed. One looks in vain for a "level playing field." The glow radiates most ardently and at highest frequencies in domains of freedom and technological creativity. In countries that have climbed highest on the spectronic ladders toward visible light, the warps and woofs of incandescence can almost be seen. Through most of the 1990s, the center of the radiance was overwhelmingly in the United States. The U.S. commanded four times as much computer power per capita as the rest of the industrial world, 75 percent of the computer networks, and 80 percent of Internet hosts. The radiance is a reflection of political and economic freedom, immigration and the spirit of enterprise. Nearly half the luminosity remains in the United States, and within the U.S., perhaps half is in California, and in California half is in the region of Silicon Valley. But the radiance is now on the move.

At the millennium, the incandescence is diffusing around the world, offering a promise of new freedom and prosperity from Santiago, Chile, to Shanghai, China. While in the midst of Europe and Africa, politicians foment entropic wars—short-circuit flares of chaos and cruelty—entrepreneurs are finding new opportunity nearly everywhere. Encircling the globe under oceans and beaming from satellites, the radiance is increasingly eroding the powers of despots and bureaucracies, powers and principalities. The crystal cathedrals of light and air are increasingly reachable anywhere on the face of the earth. To stifle links to the global communities of mind and liberty entails increasingly brutal and obvious repression and incurs rising costs of economic stagnation and retardation. Within the market space of the net, anyone anywhere can issue a petition or publication, utter a cry for help, broadcast a work of art. Anyone can create a product, launch a company, finance its growth, and spin it off into the web of trust.

Turn off your spectroscope, though, and the web disappears. It is as invisible as the life of the mind and the laws of liberty that sustain it and that it sustains. Without your Maxwellian prism, all the frequencies dissolve and—except for an infinitesimal spot of phosphorescence on your screen—disappear from sight. Although the sphere of light spans the globe and reaches up as many as twenty-three thousand miles, it appears to humans only at their screens as a single point of concentrated light. At any instant, that is all that is present on a cathode ray tube—a single spot volleying back and forth sixty times a second. It tricks your eye into seeing a full image.

Turn off your spectroscope and if you are in the wrong markets, the web may seem inconsequential. You decry "the myth of Internet traffic" as did a January 1999 issue of the estimable *Business Communications Journal.* You may call it an "American fad" as have several French executives in my presence. You may compare it to Tulipmania. For the colors of visible light are absent.

This trompe l'oeuil point of light both symbolizes and embodies the global convergence: a light of infinite dimensions concentrated into a single point racing across the screen. Moving up spectrum toward the visible domain, becoming ever dumber and more capacious, the Internet transcends geography and reduces to a single racing spot in time and space. As the commerce of the world flows into this global radiance and is distilled into this single point, it resonates with the creative work of the world. The convergence of light unleashes a fabulous efflorescence of myriad and multiplying sparks of commerce. Accommodating real time transactions, it is a market space that can absorb all the business of the new global economy. The smaller the point the more the room.

As activities move from the surface of the world and converge into the virtual pinpoint of radiance, they accelerate toward a universal resonance—the velocity of light—and converge to a universal medium. The light is both the abundance and the scarcity of the new world economy, the creative interplay of limit and infinite, the flesh and the divine.

Afterword

Beyond the Telechasm

As I was finishing this revised paperback version of *Telecosm,* our optics sage, Charlie Burger, rushed into my office with another shocking tale from the topsy-turvy world of telecosm 2002, where paradigms withered, fiber glutted, and tech shorts became the shamans of the season, strutting their stuff down the beaches of the Hamptons and on the casting couches of the money shows. In these apparent "end times," telcos were teetering on their towers of debt, with stock prices lower than Japanese interest rates, and fiber was the new "Heart of Darkness," with United Artists reissuing *Apocalypse Now* in 2001 for the telecosm era, with lugubrious lambdas and glum playmates undulating amid the napalm fumes. Doom-star analysts Ravi Suria and Susan Kalla were dirging and devaluing the entire electromagnetic spectrum, and the ITU was said to be meeting in Geneva to downgrade the speed of light.

Amid all this portentous noise, Charlie would not consign his once-in-a-lifetime scoop to so shaky a channel as a high-frequency carrier wave. Who could tell what would befall the bits as they were levered through some Ponzi scheme of JDSU lasers, Merrill Lynch junk bonds, and Nortel add/drop multiplexers, before flying down a light-wave flicker on a glass thread from Metromedia Fiber only to overrun the buffers on a line of credit and die a miserable death in the dark at a sepulchral Exodus dot-com debtor center.

So Charlie chugged into my office in person to disclose in unmodulated phonons the latest shocking news from the front: *The Internet lives. Its growth is accelerating.* Larry Roberts of Caspian, one of its prime inventors, polled the 19 leading carriers in the industry in late 2001 and learned that traffic has been increasing *faster than ever*, what he calculates as an unprecedented 3.7-fold annual pace. If it keeps up, that means a thousandfold in a little over five years. Researchers at Bell Labs essentially confirm the Roberts numbers. While all of us still wait impatiently for our information to show up, evidence from the high-end users surveyed by Keynote indicates that in the face of this 3,000-fold traffic surge since 1996, average performance has actually improved from a 12-second delay four years ago to a 3-second delay this year. The spread of cable modems and DSL may have conferred similar gains on up-market households and small businesses.

With continued growth in traffic, growth in demand for optical technology will reemerge inexorably over the next five years. The plummeting price of bandwidth and connectivity are not a problem for the paradigm; they massively confirm the telecosmic regime of abundances and scarcities.

Back in 1995—in the antediluvian age of the 14.4 kilobit-per-second modem, the two-megabit shared Ethernet local area network (LAN), and the 40-megabyte disk drive—Bob Metcalfe pointed to the ultimate source of this Internet traffic. In his 1996 book, *Packet Communication*, the Ethernet inventor and 3Com founder pointed out that the some 15 terabytes per month of Internet Protocol (IP) packets on the public network at the time amounted to a mere burble of noise beside the potential of 15 exabytes per month of Ethernet traffic then coursing through the LANs of the nation's businesses. As Metcalfe estimated, 1995 Internet traffic was just one millionth of Ethernet traffic.

At the time I had no clear notion of what an exabyte was, so I looked it up. It is 10^{18} bytes, an inconceivable vastness, best measured in LOCs. At 20 terabytes or 20 million megabytes, the LOC has found favor as a unit of measurement designating roughly the contents of the Library of Congress translated into digital form. Since a megabyte can hold the information in a book of some 400 pages, a LOC comprises about 20 million big books. An exabyte is 50 thousand LOCs, which comes to a trillion big books. The 15 Ethernet exabytes would mean more than 15 trillion big books. Imagine a tower of tomes 200 million miles high, reaching twice as far as the sun.

At the time, I was a lowlander who eked out my living by puffing self-importantly into the sails of the Net's then meager vessels of Usenet news and e-mail and bulletin-board twaddle and declaring them an armada that would soon overthrow the empire of television. Metcalfe by contrast was an Olympian inventor and entrepreneur. Shocking was his image of these kilo-LOCs looming over the frail defenses of the Internet like a vastly swollen North Sea pressing against the dikes of Holland. Metcalfe wanted to know: What if all those Ethernets then only trickling onto the Internet through PC dial-up modems seriously began to leak? What if they should burst through to mighty coaxial cables or T-1 lines and pour their contents into the then little "T-3" 45-megabit backbones of the Net? It would drown. Glug, glug, gurgle, gurgle. We would see it no more.

Ruling on Wall Street, however, has been an entirely different fear, based on the belief in a continuing glut of bandwidth that dooms the entire "next generation" of networks such as Global Crossing, Williams, Level 3 and MFN with their millions of miles of fiber. Was Metcalfe crazy when he proposed that the real threat to the Net was too little capacity rather than too much?

For gauging the current potential overhang between private networks and the public Internet, the best measure is storage capacity. Metcalfe raised his exaflood alarm in 1995. In that year, according to storage king EMC estimates, the total storage capacity of the world's computers was some 200 terabytes, including a total of 75 terabytes at 9,000 mainframe sites. In 1995, most disk drives were not connected to the Net. By 2002, most were, and total storage capacity has mounted to 10 exabytes or more. This represents 50,000-fold growth in less than a decade. In 1995, 95 percent of storage was analog (such as films, tapes, and microfiche) and unsuited for transmission on the Internet. By 2002, 95 percent of storage was digital. The onset of digital video disks (DVDs) and the dominance of CDs make movies, music, and photographs increasingly digital and Web-ready. Online storage of all the some 52 billion photos snapped annually would require 26 petabytes. As photos become digital and then reverberate their petabytes in video some 24 times a second, image traffic on the Internet will swamp everything else.

I do not use exclamation points! But let me point with an exclamatory wave of my arms to these numbers. Distributed across the globe, 10 exabytes translates into 10 billion bytes of digital information—some ten *thousand* big books of data—for every man, woman, and child on the face of the earth. Numbers so large invite skepticism. No one knows much about what the information is or why it was stored, or where it resides in what bloc, file, logical unit, disk array, rack, data center, warehouse, bitmine, cache, cave, cage, or other mnemonic device. Like most big books the data is seldom opened or read; 85 percent of it is said to be random, redundant, or obsolete. But hidden in its recesses and speleological mazes, in its caves and cages, disk farms and tape racks, are vast potential resources for an information economy.

Most of Metcalfe's floods do not now flow because missing still is the needed broadband link between enterprise exabytes and the Internet core. Late in 2001, however, help arrived from this very same Bob Metcalfe, now a venture capitalist at Polaris in Boston. In tow he had a new company, Narad Networks, which proposes to open the channels between enterprise and the Internet, unleashing the exaflood through cable TV systems, previously regarded a broadband option exclusively for homes.

Direct fiber connections to homes and small businesses are still scarce and expensive. Afflicted by regulation and the meager capacity of the twisted-pair copper medium, Digital Subscriber Lines struggle to provide sub T-1 bandwidth at typical distances. Free space optics from Terabeam offer chiefly an urban high-rise solution. Since the beginning ("Why Cable Will Win," *Forbes* 1990), my favorite residential access medium has been coaxial cable used for cable television. The reason was simple. Unlike the twisted-pair copper owned by your local telephone company, coax is inherently broadband. The potential of its usable spectrum is a hundredfold greater than the possible 50 MHz of twisted pair as upgraded for advanced Digital Subscriber Line.

Nonetheless, even with coax's inherently broadband capacity, cable's shared neighborhood lines and its upstream bandwidth cap limit most cable modem subscribers to a DSL-like 1 megabit per second, not good enough for robust business service, certainly not enough to carry Metcalfe's floods.

Putting a capacity ceiling of around a gigahertz on cable's capacity has been the inability of conventional silicon devices to convert analog frequencies above 860 MHz to digital and process them into a usable signal. Surmounting this challenge, Metcalfe's new technology from Narad transforms the huge feasible bandwidth of cable into a switched Ethernet resource, potentially bringing the huge video traffic of television onto the Internet together with petabytes of business traffic, and making cable companies the stars of the last mile. Narad Networks doubles the bandwidth of cable coax at the head end with a $10,000 box and supplies new access taps for connection to the network. Without affecting existing cable services or customers, these $200 units can be inserted at any point on the cable network to tap the new bandwidth for full duplex 100-megabit or gigabit-per-second Ethernet links.

With Narad, coaxial cable running by 65 million American homes and roughly half of American businesses will soon be able to provide 1,000 times the capacity of copper telephone lines, 60 times the capacity of a T-1 line, and 100 times that of DSL. At the same time, urban business customers are turning increasingly to gigabit Ethernet services from such providers as Yipes and Cogent. For rural areas, satellite companies such as Hughes and Echostar will tend to dominate because, despite limited total capacity, they offer spanwidth: three geosynchronous birds can cover the entire globe. Promising an increasing spread of digital voice services as well, this new competition from cable, glass, and air will drive the telephone companies to accelerate their move toward fiber to the curb and household.

The telecosmic collapse, as painful as it has been, is mainly a monetary event. The financial world is undergoing an acute deflation, which puts every indebted company—and country—through the wringer, as it must pay back its loans and bonds in dollars that have appreciated against gold and commodities by at least 40 percent in five years.

The crash in prices has concealed the accelerating advance of real-world optical technology and Internet traffic. Combine the traffic data with SG Cowen data on carrier capital-equipment expenditures, and you can calculate a drop in the cost of bandwidth of 58 percent per year, or roughly a hundredfold decline over the last 5 years. Fresh from deploying and lighting up some 42 OC-192-ready channels for Sphera Corporation in a record 14 days in mid-2001, Rohit Sharma, chief technology officer of ONI (ONIS), believes that this rate of advance is sharply accelerating. ONI will introduce WDM gear offering a tenfold increase in cost-effectiveness by the end of 2002, a rate of advance equivalent to the 79 percent decline in bandwidth costs estimated in 2001.

What has happened is a manifestation of the triple Moore's law pace of bandwidth advance. With a near doubling of capital outlays between 1998 and 2000 and a continuation of the trend expected, the CapEx (capital expenditures) slump in 2001 was concussive to the industry. But capital is not money; it is capability. Adjusted for increasing cost-effectiveness, actual bandwidth deployment or *real* capital spending continues to soar. At the carrier level, price elasticity of demand for bandwidth continues to register at a rate of between four and six (a 50 percent price drop yields a 200 to 300 percent increase of quantity demanded). The price of a unit of long-distance bandwidth (a DSO-mile-month in telco talk) has dropped 500-fold since 1996, while traffic has increased 3,000-fold, yielding a crude elasticity number of six.

These reports signify a pace of advance in transport and switching technology that may be fairly described as stupendous. Charlie Burger calculates that to sustain 2001 Internet traffic with the communications technologies of five years ago would have cost *$39 trillion* or four full years of GDP. Dealing at once with a tsunami of new traffic and crunch of economic deflation, the bandwidth stars of the telecosm have performed a job far more in line with the stock prices of the so-called "bubble" of the year 2000 than of the crash of 2001.

This book celebrates these feats of bandwidth abundance. The collapse of the price per bit of bandwidth is exactly the consummation that we have been seeking and predicting and explaining for the last ten years. In the telecosmic scheme of abundances and scarcities, bandwidth is indeed the canonical abundance. But as the renaissance visionary David Gelernter of Yale and Mirror Worlds gently reminded me in a review and in a recent interview, bandwidth is only one side of the solution. Slowly, but rather surely, the information industry is learning to waste bandwidth. However, as Gelernter observed, "Industry has yet to learn how to squander *storage* creatively."

Throughout most of the computer age, storage of all kinds has been oppressively scarce. Until the early 1990s, hard drives advanced far more slowly than Moore's law, with real densities increasing at a rate of only 27 percent per year. All archival storage had to go on tapes and much of it remained analog. Then came the explosion: a set of synergies in platter quality and flatness; semiconductor thin-film fabrication; magneto-resistive drive heads; fast, accurate servomotors; digital-signal processing for bit detection; and related software, all rushing forward in parallel to unleash a miracle in hard drives.

Doubling every year—compared to a Moore's law cycle of 18 months—a high-end PC disk drive moved from 100 megabytes in the year 1992 to a scheduled 100 gigabytes this year. In 1992, a 40 megabyte drive cost $150. In 2002, $151 gives you a 60 gigabyte GXP from IBM, which is shortly to launch a 120 gigabyte model, likely available to early adopters for $400. The 60 GXP represents a 1,500-fold drop in price per bit in a decade.

Optical bandwidth makes this storage accessible across the planet. But it cannot surpass the lightspeed limit. Different distances and disruptive delays afflict the current Internet with latency and jitter that frustrates streaming audio and video and undermines the computer as a replacement for the TV. A rule of thumb in computer design ordains that bandwidth problems are solved in hardware, latency problems in software. To improve performance in latency and response is chiefly a problem of developing parallel software to tap the global scatter of abundant resources in storewidth. A 120 gigabyte drive can hold 60 full-length MPEG2 movies, or 24,000 MP3 songs, or 100,000 high-resolution digital photos, or scores of downloaded TV programs and home videos. But this abundance of content is valuable chiefly to the degree that users can share it. As home storage mounts, a peer-to-peer paradigm—computers in the home and business-sharing resources across the network without a top-down structure—is inevitable.

This model now collides, however, with perplexing issues of copyright and intellectual property. Both in court and in the marketplace, the current strategies of the music industry are doomed to failure. In early 2002, Napster's free successors were handling close to double Napster's peak 9.6 petabytes of February 2001. Most of the new peer-to-peer file-sharing systems are merely turbo-coded versions of ordinary file-transfer software. To stop them, the music industry must outlaw the basic structure of the Internet.

When your product is stolen by thieves, you have a police problem. When it is stolen by millions of honest customers, you have a marketing problem. At present, otherwise honest people download daily 150 million songs free over the Internet. That means 50 billion songs a year, the equivalent of 3 billion CDs of music, or more than three times the 850 million real CDs sold annually. The "Napster" effect has roughly quadrupled the number of songs used in a year, expanded dramatically the range of music heard, and potentially enlarged the industry. Now the challenge is to make it into a viable business.

The music industry's market responses to Napster—Press-Play, from Sony and Universal, and Music.Net from EMI, Warner Brothers, and Bertelsmann—are trying to compete with a free service by offering less. They bar the sharing of music files, lists, and combinations that are the focus of current music users on the Net.

Offering a more promising scheme is Rob Reid, author in 1997 of the first definitive book on Internet commerce, *Architects of the Net*. Now aiming to become one himself, he has launched a company called Listen.com. It has created a system for distributing songs that both accommodates the current users and lays the foundations for a heavily profitable music business on the Net. Transmitting music files in unplayable form called residual audio, Listen.com makes no move to inhibit the use of them until they are played. At that point, Listen.com transmits a secure, real-time activation strip called essential audio containing just 2 percent of the volume. Reid's strategy is to allow the user to do whatever he likes with the file—storing, listing, filing, exchanging it—until he actually plays the item. At that point he must qualify as a subscriber or a micro-payer in order to receive the real-time, secure activation strip that renders the material playable. Designed for a world of storage and bandwidth abundance, where the prime impulse is sharing experiences, Reid's model can spur the Net to new heights and create a new arena of profitable commerce.

In 2005, just three years from now, will begin the terabyte era, when a few hundred dollars will get you a terabyte drive, which could hold 480 films, or 200,000 songs, and so on, in digital form. And that is in just one household or office desktop. Put them in 100 million households around the globe five years hence and you have 10 exabytes of storage just in people's homes. This is storage abundance with nary a storewidth challenge in sight. Over the short hop between your disk and your display, or your basement and your entertainment room, these songs and images, films and games will play out utterly without latency or jitter.

Napster-style users do not worry about disrupted links. Nor do TiVO or Replay television customers, who download their favorite programs into what now seem meager 40 gigabyte drives. Nor will the millions of users of multigigabyte home storage. As Nicholas Negroponte predicted eight years ago, the bulk of Internet transmission will be store-and-forward. The medium for store and forward is peer-to-peer. Today peer-to-peer is the dominant way of transferring music files of 3 to 5 megabytes; tomorrow, peer- to-peer will prevail in the transmission of 2 to 5 gigabyte video files.

As the disk drive industry continues its advances, the Listen.com strategy naturally morphs into Watch.com, usable for films, games, and multimedia. Extended to video, the chief difference is the hundredfold greater size of the files. While a CD-quality song averages around a megabyte a minute, a high-resolution video consumes about 100 megabytes a minute. A world of images will entail both storage and bandwidth roughly a hundred times greater than today's. At the current pace, that will happen in less than five years.

Encompassing this peer-to-peer regime will be an infrastructure of servers, archival storage, data-mining facilities, and other technologies that will originate the product, protect its intellectual property, maintain archival libraries, and support and complement its use. Listen.com is deploying a 5-terabyte server farm to deliver songs and strips to its subscribers.

Parallelism and peer-to-peer processing and computing technologies are everywhere changing the face of storewidth. Parallelism is what makes Google's search engine sort through 3.6 billion plus web pages and bang out the search results in typically less than one-fifth of a second. Google uses some 10,000 Linux PC servers in parallel, each linked to Maxtor 80 gigabyte drives. In the past, in order to pump gigabytes of data onto the Net, you needed a $50,000 server with multiple ports and processors, queues, "threads," and buffers, and with costly specialized disk facilities and databases. For many crucial functions, including rapid simultaneous downloads from high traffic sites, such architectures remain indispensable and will proliferate with Metcalfe's flood. But a Bear Stearns study on Internet infrastructure by analysts Chris Kwak and Robert Fagin calculates that in January 2001, Napster's 65 million registered customers unleashed a tsunami of 9 petabytes of MP3 digital music onto the Net, between a quarter and a half of all Internet traffic that month, with no such centralized functions at all. If Napster were built on a centralized model, it would have had to purchase over 5,000 F840 6-terabyte Enterprise Filers from Network Appliance, at a cost of some $666 million to hold all the songs, and then purchase bandwidth at a cost of $6.7 million per month. But Napster centrally commanded only a small disk space for pointers, addresses, and song lists; it used the disk drives of its millions of customers as its storage facilities and pays not $666 million, but nothing at all.

The storewidth paradigm dictates wasting the increasingly free resources of bandwidth and storage to bypass the slow, costly costive snarls of bits and wires, routers and servers, silicon and copper, switches and buffers, queues and protocols that currently populate the Internet. Overwhelming, all tricky, customized solutions—Ethernet moving to 10 gigabits per second, Infiniband unleashing fast servers on industry-standard platforms, cheap storage multiplying its capacity into the terabytes per drive—all will advance on their parallel paths.

Rather than optimizing our own processors to deliver huge amounts of delicate data across the Net, these parallel systems allow us to exploit the abundance already in place: processors that are free because they are already there in the PCs of the world. Making these PCs yet more valuable is a new method of sending flawless files through noisy channels. Based on a new "holographic" algorithm invented at Berkeley by mathematician Mike Luby, a company called Digital Fountain offers a system for downloading any file perfectly regardless of interruptions, sequence, jitter, or lost packets, with just 5 percent overhead. The secret is yet another form of parallelism. All packets are created equal, and as long as the recipient gets a full complement plus 5 percent, the file is flawlessly transferred. Rather than sending the content itself, Digital Fountain puts it into a mathematical blender that homogenizes the packets. While the laws of entropy prohibit reseparating scrambled eggs, the laws of mathematics—and Luby's algorithm—allow the receiver to reconstitute perfectly the original file from the transmitted metadata. Launched with support from Garage.com and yours truly and backed by Sony and Cisco and an array of top line venture capital, Digital Fountain is transforming the parameters of file transfer on the Net.

Consider the storewidth dilemma faced by Siebel Corporation developing customer response management software in San Mateo and Ireland and shipping the "builds" across the globe every night. Beginning with a T-1 line, the transfer took seven hours at a pace of 500 kbps. A seven-hour time-change made this process intolerably late. Hey, this is the age of bandwidth, the Siebel people figured, "throw bandwidth at the problem." So over a period of weeks, they acquired a 30-megabit-per-second (fractional E-3) line to Ireland and sent off the nightly build. Huh? The transmission still traveled at 500 kbps and took seven hours. Ignoring speed of light latency, which consumed 200 milliseconds for every roundtrip TCP acknowledgment and required to-and-fro retransmission for every lost packet, Siebel was trying to solve a storewidth problem with bandwidth.

Siebel discovered that Digital Fountain servers together with less than a megabyte of client software could remove the latency. Gone were all the delays and back channel acks and reacts. The result was a fivefold acceleration of the transfer. With all packets equal, lost packets do not entail retransmission, merely a proportionate extension of the flow. If faster transfer is needed, multiple streams can be sent in parallel with no change in the file and no need to get them in order at the other end. If you want to do streaming audio or video, you break up the file into segments and send each of them to be buffered at the other end. If you want to filter a file through the troposphere from a geosynchronous satellite or bounce it around the globe through 25 routers that routinely relieve congestion by dropping packets, you need no back channel signals or retransmissions to assure a perfect transfer.

Over the next five years or so, the Internet core will turn into an all-optical fibersphere which will convey data on massively parallel wavelengths without dropped packets and multiple hops, jitter and jumble. However, wireless, satellite, and other noisy last-mile channels assure that even then, the Digital Fountain or similar technology affords a cheap and effective way to get the most out of the storewidth jungle.

The ultimate parallel resource of the telecosm is human vision and cognition. All visual reality is coded as oscillations and waves of neural excitation that wash back and forth across the brain and enable identification of edges and contrasts in an image. The 6 million cones and 120 million rods in the human retina initiate this process, enabling the eyes to accept a fabulous 15 orders of magnitude (a factor of a thousand trillion) of dynamic range or brightness levels. But with 500-micron spacing between the cones and 30-millisecond latency in the rhodopsin molecules that detect and convert light, human eyes can receive no more than one gigabit-a-second of information. Moreover, the 126 million rods and cones link to the optic nerve through the only 1 million ganglion cells—a compression ratio of 126 to one—that reduces the apparent flow to just 7 megabits a second. But it gets worse. Sight itself—the actual ability of the brain to absorb and interpret words or images—remains intractably at around 25 bits a second.

The brain commands a broadband antenna (the eyes) and a slow, narrowband, but profoundly intuitive processor, that is yet to be understood by science. The key to creating computational capacity to complement the brain is the development of computer architectures that exploit what David Nolte portentously calls "the parallel advantage of light and image." At the heart of such architectures will be images that can serve as programs to process other images. Since images are inherently analog, the ultimate light computer will be an analog machine, working with interference, refraction, and defraction among intersecting colors of light with different phases and amplitudes. Digital is great if you don't know what you are doing and need to program it later. But if you know what you are doing—identifying a preset pattern, face in the crowd, code in a congested cell, missile in a crowded sky, or WDM array of closely packed channels—analog optics is far faster and more capacious. If you can use the image as a program to process other images, you can do massively parallel processing in real time. By overlaying one image over another, you can cancel all the redundant information and leave only the desired deltas or deviations.

The ultimate promise of the telecosm is to integrate a global optical Internet with billions of light processors working in parallel to accomplish most of the business of the world. Fulfilling that promise will be new waves of innovation, undercutting costs and sweeping away even apparently ascendant technologies. As prices plummet, vendors of high-priced spreads will howl in pain, and analysts will once again denounce canonical abundances as "gluts," as if the world economy could have too much intelligence or too much capacity to communicate it. As always, the new abundances will be created by men who cannot themselves fully imagine their purpose. "Leap before you look" is a law of enterprise. You cannot see anything truly new from an old place. You must ride the light.

APPENDIX A • A List of Telecosm Players

ASCENDANT TECHNOLOGY	COMPANY (SYMBOL)	52 WEEK RANGE*	MARKET CAP*
FIBER OPTICS			
Optical Fiber, Photonic Components	Corning (GLW)	12.60–113.33	14.6B
Wave Division Multiplexing (WDM) Components	JDS Uniphase (JDSU)	7.90–128.00	12.2B
Adaptive Photonic Processors	Avanex (AVNX)	6.82–161.38	452.7M
All-Optical Cross-Connects, Test Equipment	Agilent (A)	25.00–68.00	13.1B
Crystal-Based WDM and Optical Switching	Chorum (private)		
WDM Metro Systems	ONI (ONIS)	15.75–114.75	3.2B
WDM Systems, Raman	Corvis (CORV)	3.00–114.75	1.4B
Metro Semiconductor Optical Amplifiers	Genoa (private)		
Optical Processors	Essex (ESEX.OB)	1.38–6.70	24.2M
LAST MILE			
Cable Modem Chipsets, Broadband ICs	Broadcom (BRCM)	20.88–274.75	11.4B
S-CDMA Cable Modems	Terayon (TERN)	2.36–61.38	441.7M
Linear Power Amplifiers, Broadband Modems	Conexant (CNXT)	6.90–54.94	2.4B
Broadband Wireless Access, Network Software	Soma Networks (private)		
Broadband Wireless Access, Network Software	Nayad Networks (private)		
WIRELESS			
Satellite Technology	Loral (LOR)	1.03–8.50	710.2M
Low Earth Orbit Satellite (LEOS) Wireless Transmission	Globalstar (GSTRF)	0.25–14.19	42.7M
Code Division Multiple Access (CDMA) Chips, Phones	Qualcomm (QCOM)	42.75–107.81	48.1B
Nationwide CDMA Wireless Network	Sprint (PCS)	15.72–56.06	24.3B
CDMA Handsets and Broadband Innovation	Motorola (MOT)	10.50–37.25	41.2B
Wireless System Construction and Management	Wireless Facilities (WFII)	3.31–82.69	361.0M
Internet Backbone and Broadband Wireless Access	WorldCom (WCOM)	12.50–38.38	41.3B
GLOBAL NETWORK			
Metropolitan Fiber Optic Networks	Metromedia (MFNX)	0.49–40.19	528.9M
Regional Broadband Fiber Optic Network	NEON (NOPT)	3.40–50.13	101.4M
National Lambda Circuit Sales	Broadwing (BRW)	15.40–30.00	5.3B
STOREWIDTH			
Java Programming Language, Internet Servers	Sun Microsystems (SUNW)	12.85–64.69	53.1B
Network Storage and Caching Solutions	Mirror Image (XLA)	2.49–29.00	291.8M
Remote Storewidth Services	StorageNetworks (STOR)	4.25–141.00	601.9M
Hardware-centric Networked Storage	BlueArc (private)		
Virtual Private Networks, Encrypted Internet File Sharing	Mangosoft (MNGX.OB)	0.53–12.75	28.9M

ASCENDANT TECHNOLOGY	COMPANY (SYMBOL)	52 WEEK RANGE*	MARKET CAP*
MICROCOSM			
Analog, Digital, and Mixed Signal Processors	Analog Devices (ADI)	30.50–103.00	16.6B
Silicon Germanium (SiGe) Based Photonic Devices	Applied Micro Circuits (AMCC)	11.25–109.75	5.2B
Programming Logic, SiGe, Single-Chip Systems	Atmel (ATML)	7.63–21.94	4.6B
Single-Chip ASIC Systems, CDMA Chip Sets	LSI Logic (LSI)	13.65–40.75	7.9B
Single-Chip Systems, Silicon Germanium (SiGe) Chips	National Semiconductor (NSM)	17.13–47.94	5.6B
Analog, Digital, Mixed Signal Processors, Micromirrors	Texas Instruments (TXN)	26.26–71.00	59.8B
Field Programmable Gate Arrays (FPGAs)	Xilinx (XLNX)	29.79–92.50	13.3B
Seven Layer Network Processors	EZchip (LNOP)	3.69–38.44	40.9M
Network Chips and Lightwave MEMS	Cypress Semiconductor (CY)	13.72–49.94	3.4B
Field Programmable Gate Arrays (FPGAs)	Altera (ALTR)	18.81–67.13	11.6B

*July 31, 2001

Appendix B

Nine Stars of the Telecosm

COMPANY/MARKET SYMBOL
ASCENDANT TECHNOLOGY

Altera (ALTR)
Field Programmable Gate Arrays (FPGAs)

Avanex (AVNX)
Adaptive Optics

Broadcom (BRCM)
Single-Chip Broadband Data Transmission

Corning (GLW)
Optical Fiber, Photonic Components

Corvis (CORV)
WDM Systems, Raman

JDS Uniphase Corporation (JDSU)
Broadband Optical Networking Components

ONI (ONIS)
WDM Metro Systems

Qualcomm Incorporated (QCOM)
Broadband Wireless Technology, Equipment

Texas Instruments (TXN)
Mircrochips: Digital Signal/Mixed Signal Processing

Altera

Altera (ALTR), founded in 1983 by Bob Hartmann and others, has built a billion-dollar business with a narrow focus on programmable-logic devices (PLDs). PLDs are general-purpose chips that can be "programmed" to implement custom-digital circuits. Altera and Xilinx dominate the programmable-logic market. PLDs support today's Telecosm, and they will be key components in tethered devices for the fibersphere's future. Altera has been growing as fast as it can to stay abreast of customer demand. It has built its business on prototyping and on displacing application-specific integrated circuits (ASICs). Today, Altera is on the leading edge of a long-running wave of success as PLDs displace ASICs. Moore's law says the number of transistors on a chip doubles every eighteen months. High-end chips get bigger and faster. Also as a consequence of Moore's law, chips with a fixed number of transistors get smaller, cheaper, and faster over time. Moore's law is either your friend or it is your enemy, but it is fickle and will change sides.

Think of PLDs and ASICs as suppliers of circuits. The supply of circuits is growing at a Moore's law rate. The supply of circuits is growing faster than the demand for circuits. PLDs, because of the overhead in transistors that support their personalization, are well below ASICs in the number of usable circuits that they can supply. But, because both PLDs and ASICs grow rapidly in the number of circuits they can supply, PLDs are usurping market segments held by ASICs.

Historically, PLDs have been slow, they have been short on circuit capacity, and they have been expensive. PLDs have an advantage in flexibility; the designer easily corrects errors in a PLD-based design, but the ASIC has to be scrapped. Designers used PLDs to build and to debug the first implementations of circuits that would later be cast as ASICs. An ASIC is faster than a PLD and, in high-volume production, it is cheaper. The ASIC's cost has two components: a high fixed cost and a low variable cost. The fixed cost is for the design of its mask set for manufacturing. The variable cost is the per-unit cost for production. Since the PLD is a general-purpose chip, the cost of its mask set is amortized into its per-unit production cost. Each year PLDs become faster, they increase circuit capacity, and, at least at the low end, they become cheaper. The demand for more circuit capacity and more performance in ASICs is rising each year, but Moore's law is driving the supply of circuit capacity and performance in PLDs faster. As a result, Altera's PLDs are displacing ASICs at the low end of circuit capacity and performance.

That's the way the PLD business has worked since it started: it's been prototyping and it's been eating into the low end of the ASIC market. It's a great business and it will continue to grow. More potent than the PLD's advantages in prototyping and in competing with ASICs, however, are its advantages in such larger markets as microprocessors and digital-signal processors (DSPs).

Since the commercial introduction of the microprocessor by Intel in 1971, Moore's law has been working for the microprocessor as it displaced custom integrated circuit (IC) macro functions and rode semiconductor-fabrication improvements to billion-unit volumes. The microprocessor is a general-purpose chip that brings the computer's problem-solving method (programming) to the hardware domain. The microprocessor displaced IC macro functions because it was cheaper and because it had adequate performance. But, because of its programming languages, compilers, operating systems, and general-purpose instructions, the microprocessor is not as efficient as direct-hardware implementations.

For thirty years, the microprocessor has compensated for its lack of efficiency by improving its performance. The microprocessor's performance improves with increasing complexity (more transistors) and with rising clock frequency. But power dissipation also increases with increasing complexity and with rising clock frequency. As the computational requirements of applications have become more demanding, the microprocessor's clock frequency has risen above 1,000 megahertz (MHz) and its power dissipation has climbed to the range of 50 to 60 watts. As long as power was free and efficiency wasn't important, Moore's law worked for the microprocessor. But Moore's law is about to change sides. As power and efficiency become important, it will switch sides to work for the PLD and it will become the enemy of the microprocessor.

The microprocessor and the DSP are general-purpose components. They displaced direct-hardware implementations with programmed solutions, even though they are inefficient and slow in comparison with a direct-hardware implementation. The microprocessor and the DSP displaced direct-hardware implementations because they had adequate performance, they increased the engineer's productivity, and they were cheaper. They were cheaper by virtue of high-volume production of a limited number of standard components applied across a vast range of applications.

Moore's law has given the microprocessor and the DSP a long honeymoon. This couple has been riding a wave of speed and capability improvements that has let them dominate the application space. As a result, the engineering community, educational system, and development tools look first to microprocessor-based solutions. The low efficiency of their solutions hasn't been an issue because performance has been adequate and power has been free. Now power and efficiency are becoming important. For mobile applications, power and efficiency are obviously important, but they are becoming important in tethered applications too. Ask the manager of a server farm if it's easy to supply 5 kilowatts per rack to several hundred racks of computers. The average server farm in Silicon Valley asks the power company to deliver twelve megawatts (that's fifteen to twenty times what a typical commercial office building uses).

Direct solutions implement the application's algorithms in hardware, rendering engineering equations representing the problem solution directly into logic circuits. Direct solutions offer performance and efficiency, but require considerable engineering effort. Microprocessor-based solutions program the problem on a computer's instruction set. It is inefficient and indirect, but programming the solution increases the engineer's productivity. PLDs lie in the middle between direct-hardware solutions and programmed solutions. They are more efficient than programmed solutions but require more engineering effort.

The PLD has advantages both in the directness of its circuit implementation and in its potential for parallel logic. The microprocessor is limited to the resources it has when manufactured. If the problem could use fifty multipliers, the PLD can have them. With fifty multipliers, the PLD might solve at 50 kHz a problem that would require 1,200 MHz from the microprocessor. The PLD's parallel implementation is more efficient than the microprocessor's serial instructions and its lower clock frequency conserves power. Altera's PLDs displace ASICs because they consolidate one-of-a-kind designs into general-purpose chips. They displace microprocessors and DSPs because they offer more efficient, more direct solutions that conserve power. Altera's business fits the Telecosm vision. Altera is among the companies poised to change the world because Moore's law is on its side, while it is working against ASICs and against microprocessors and DSPs.

Avanex

The name Avanex (AVNX) will be familiar to most of you. Some of you who bought its shares in the aftermath of its $14 billion IPO may be ruing the day you learned it. But you can be assured that in those buildings, replete with some 450 animated Asian workers, you can find some of the most creative minds in all of optics and the single most promising new product in the industry, the Avanex PowerMux, or symmetrical multiplexer-demultiplexer.

This device performs the vital function of taking bit streams from many wavelength sources, combining these wavelengths at one end, and sending them simultaneously down a single fiber. At the other end, it takes these same wavelengths and separates them, along with their bit streams, so they can be sent to their destinations. Unlike most mux-demux devices, the Avanex processor is scalable and adaptable. It can handle virtually any bit rate, any number of wavelengths, and any spacing between wavelengths. In other words, unlike the products of its competitors, it is not restricted to a fixed set of capabilities that maximize only the gross capacity of the fiber. Instead, PowerMux maximizes the flexibility and adaptability of the network. The principle behind the PowerMux could re-create today's network, consummating the silent revolution launched by WDM, reversing the thirty-year progress of packet switching, restoring the elegant luxury of circuits, but now enriched many millionfold by the power of light.

The key to every new economic era is a canonical abundance. Companies prevail by wasting it. During the transistor age, companies that husbanded transistors—designing them carefully and implanting each one at the optimal spot on the microchip—lost out to companies that wasted transistors. Rather than honing single transistors to run ever faster, the winners accelerated chip design to launch new products ever faster and squandered transistors as if there were no tomorrow.

Today the ascendant technology is optics and the canonical abundance is bandwidth. Companies focused on jamming more and more information packets down a single-lane bit stream—as if there were no bandwidth to spare—will lose to companies that waste bandwidth in order to build capacious multilane highways with each lane running well below capacity.

For lanes, think wavelengths or lambdas. The networks of the future will rarely need light paths that can bear the Net traffic of entire cities on a single beam. What the new networks will soon require are millions of addressable colors of infrared light. Each one will constitute a potential circuit connection between one terminal and another, just as your telephone creates a circuit connection between one user and another.

This mandate is difficult for most experts in the industry to grasp because it seems to reverse every important advance in networking over the past three decades. The leaders of telephony came of age years ago in an environment of bandwidth abundance, at least when measured against the modest demands of voice. They wasted bandwidth as a matter of course. Most of the capacity of a telephone network lay fallow more than 95 percent of the time as people used their phones an average of twenty minutes a day. In a world of bandwidth abundance, circuit switching—connecting the two parties over a line devoted entirely to their call—made sense.

As the Internet rose and data became dominant, however, computers remained on line for many hours at a time. Even as absolute bandwidth soared, it grew more scarce relative to demand, and the phone companies were beset with complaints of bandwidth bottlenecks. Changing their entire approach to network economics, the masters of telecom in the 1990s had to learn how to economize on bandwidth. From Bell Labs, they learned the secrets of statistical multiplexing, combining many calls onto a single long-distance connection. Then from the Internet, they laboriously learned the rules of packet switching, cutting up every message into many packets, each bearing a separate address. While a circuit-switched phone network sets up the call in hundreds of milliseconds, a packet-switched network functions like a multimegahertz post office. The envelopes are switched not in minutes or even milliseconds but in microseconds.

Finally, telephone and data network engineers have definitively learned the superiority of packet switching. The new optics is turning the entire world of networks upside down once again. Wavelength-division multiplexing, which sends many colors of light down a single fiber thread, is ushering in a tide of fabulous bandwidth abundance. Back to the drawing board one more time, folks. In a world of bandwidth abundance, bandwidth-wasting circuits become ideal once again. Rather than economizing on bandwidth by chopping everything into packets and multiplexing them into time slots, the mandate is to waste bandwidth. As in the old telephone system, the best approach is circuits. In this case, the system software sets up wavelength circuits between terminals at the edge of the fiber network. At first these terminals will be giant Juniper or Cisco routers in Internet hubs and data warehouses, where the wavelengths are finally converted back into packets or launched into the fiber "cloud." But the reach of wavelength circuits will steadily expand into metropolitan networks, across corporate campuses, and finally into enterprises and even neighborhoods. Many of the giant routers will go away.

At the helm of Avanex is a thirty-eight-year-old with more than thirty-six optical patents named Simon Cao, one of the leading figures in the history of optics. Having smoothly disbanded the company's distracting software arm, Walter Alessandrini is a capable and decisive CEO. But it is Cao who creates and manages the technology. In 1993 at E-Tek Dynamics (now owned by JDS Uniphase), Cao helped develop the first WDM system, a two-channel affair based on a thin-film filter he engineered. Working with Pirelli of Italy, he then proceeded to create the key components for a four-channel system based on the first EDFAs. An amplifier that rendered WDM efficient by boosting all the wavelengths at once, the EDFA was enabled by Cao's ingenious isolators. These devices combine polarization effects with filters to keep the light going in one direction rather than being partly reflected. In 1996, still at E-Tek, Cao contributed crucial components to Alcatel's thirty-two-wavelength system.

In 1997, Cao left E-Tek to help found Avanex and launch his new inventions, led by the PowerMux. Until the Avanex device, optical suppliers, such as Lucent and JDS Uniphase, created multiplexing gear using devices called gratings that operate in the spacial domain. The new director of research and development at Avanex is Giovanni Barbarossa, who worked on these gratings at Lucent and came to see their limitations. These have to manipulate wavelengths differing by billionths of a meter (nanometers) in size. By contrast, the PowerMux operates in the frequency domain. It manipulates colors of infrared light billions of cycles a second apart. For example, a wavelength of 1,550 nanometers (1.55 thousandths of a millimeter), typical in WDM systems, travels through a point in space at a frequency of approximately 194,000 gigahertz or 194,000 billion cycles per second. For longer wavelengths, the frequency decreases, since the number of cycles that can pass through a point every second at the constant speed of light will decrease.

Shorter wavelengths have higher frequencies. At 1,550 nanometers, a mere 1-nanometer change in wavelength results in a change in frequency of 133 gigahertz—133 billion cycles per second. Clearly, it is easier to separate frequencies billions of cycles a second apart than their corresponding wavelengths only small fractions of nanometers apart.

Once a particular spacing is set, the PowerMux is a periodic processor that automatically sets up the channels at the correct gigahertz spacing. The tighter the spacing, the more wavelengths you can fit on a fiber. At OFC Avanex demonstrated 800 channels spaced 12.5 gigahertz, or only 0.1 nanometer, apart. This was accomplished in the frequency domain. Barbarossa doubts that other companies can achieve such spacing as reliably and robustly in the nanometer domain of wavelengths.

The PowerMux mirrors' distances and reflectivities can be changed in minutes to accommodate or tune to different sets of wavelengths. This tuning speed, though slow in the world of circuit switches, is still a vast improvement over nontunable fiber Bragg gratings. And Simon Cao envisages a time when the PowerMux can be dynamically tuned in thousandths of a second, or milliseconds. Experts from the packet-switching world—where millions of packets must be processed in a single second—will disdain any device that switches "only" in milliseconds. But millisecond switching is fine for setting up new circuits in an all-optical network based on lambdas.

This is Simon Cao's paradigmatic dream, a network of optical highways that does not have to be switched at all in the usual sense. Wasting bandwidth as a replacement for switching, he foresees networks in which PowerMuxes sort and shape millions of lambdas through the network. Meanwhile, tunable lasers and photodetectors do the work of creating the wavelength "phone numbers." In essence, it is a return to the old topology of telco circuits. But in Cao's world, each of the wavelengths can carry the multigigabit-per-second streams associated with high-resolution images like those now seen in digital HDTV. Circuit-based video teleconferencing can become virtually as vivid as being there.

Already joining in Cao's vision is WorldCom, which has begun ordering PowerMuxes for long-haul applications. Lined up right behind WorldCom are Nortel and Qtera, Corning, Sycamore, Siemens, Sumitomo, and Fujitsu, all in various stages of ordering the PowerMux.

Under Cao's technical leadership, Avanex plans to be ready for rapid change. So far in his career, Cao has been able to meet every challenge of WDM with new creativity and insight. I would bet on him for the future.

Broadcom

Broadcom (BRCM) is the dominant chip designer for the bandwidth explosion in the cable industry. The company, offered to the public in April 1998 at $24 a share, designs, develops, and markets integrated circuits (ICs) for broadband markets including cable settop boxes, cable modems, high-speed networking (Ethernet/gigabit Ethernet devices), satellite and terrestrial digital transmission, and digital subscriber line (*x*DSL) applications.

Broadcom focuses on four important technologies to supply the needs of these markets: communications systems algorithms and protocols, advanced digital signal processing design, advanced cell library development, and high-performance analog and mixed-signal circuit design.

An important competitive advantage for Broadcom is the custom design of chips. The company develops integrated circuit products for incorporation into its customers' equipment. The company is able to leverage its multidisciplinary expertise in system architecture, physical layer, and integrated circuit design to create faster, more highly integrated systems with lower power requirements. Broadcom has an impressive list of customers, including 3Com, Nortel Networks, Cisco Systems, General Instrument, Motorola, and Scientific-Atlanta.

Broadcom and MIPS Technologies entered into a strategic alliance in August 1998 that will leverage the high-performance MIPS RISC (reduced instruction set computing) processors in Broadcom's digital communications designs. The MIPS architecture provides Broadcom with the processor element needed to create complete systems-on-a-chip.

Broadcom has established itself as the leader in cable modem chip architecture, and its design has become the industry standard. The company has announced a new single chip design, the BCM3300, which provides the first implementation of extensions to DOCSIS (data over cable service interface specification) for advanced quality of service (QoS), which enables high-quality voice transmissions. This chip has constant bit rate capability and supports services such as cable telephony and videoconferencing over a cable network. It also supports "push services," such as stock tickers, news, and sports scores over cable networks. The chip integrates the functions of modulator, demodulator, and media access control chips in one.

In October 1998, the chip was customized to be compliant with the newly approved ITU (International Telecommunications Union) cable modem standard, and will provide European cable operators with interoperable cable modems. The BCM3300 can also be used in settop boxes and external network interface units.

In November 1999, Broadcom and Terayon were selected by CableLabs, the cable industry's R&D/standards group, to create a new advanced cable modem standard based on Terayon's S-CDMA (synchronous code division multiple access) and Broadcom's Advanced Frequency Agile TDMA (time division multiple access) in which the frequency range of the transmission channel is divided into narrower bands, each of which can carry a different TDMA signal. This direct involvement in development of cable standards further solidifies Broadcom's industry-leading position.

Using its scalable *x*DSL transceiver chip, Broadcom has introduced a new VDSL (very high bit rate asynchronous digital subscriber line) design which supports bit rates up to 26 Mbps. This rate will support the most demanding multimedia transmissions such as video-on-demand, videoconferencing, and digital TV over standard copper twisted-pair phone lines.

The company announced in October 1998 a highly integrated, low-power, and low-cost Fast Ethernet switching innovation, which reduces both system cost and board space by 50 percent. The design is plug and play, scalable, and automatically configures to any network topology. Its low cost and ease-of-use make it an ideal solution for small offices and start-up businesses.

This company has rocketed off the pad from its IPO in 1997 and will be a major player in a number of telecosm technologies well into the coming century.

Corning

Free of entanglement with the established base of electronic and optoelectronic telecom gear, Corning has been moving aggressively into telecommunications. Since 1997 it has sloughed off its breast implants, chucked its specimen bottles, thrown its famous Corningware glass into the hands of a Borden affiliate for $603 million, and emerged as a macho photonic giant. Acquiring the Optical Corporation of America (OCA) that same year while establishing a St. Petersburg optics laboratory with thirty top Russian scientists and expanding by 60 percent its domestic R&D, Corning has gone on to become, along with JDS Uniphase, our favorite broad-based components company.

The historic originator of fiber optics, Corning scientist Frank Hyde in 1934 invented fused silica, the first fiber pure enough to be used in communications. In 1978 Corning used this process to create the first single-mode fiber (still dominant today) in volume. For most of the last three decades it has been the most salient company producing fiber-optic line.

Corning's world-leading share of the optical fiber market has blossomed to 40 percent globally and to 50 percent in North America, far ahead of second-place Lucent. Corning has been a technological leader in this market; its large effective area fibers (LEAF), including an undersea LEAF, support the latest WDM systems numbering in the hundreds of channels and have been adopted by major networks for their nationwide fiber buildouts. Looking ahead, the company promises ever-improving hybrid fibers bearing dispersion profiles and other properties finely tuned to the sensitivities of different networks, major achievements in a world where tolerances are measured in tenths of microns and factories account for the curvature of the earth.

Today, having integrated beyond fiber to become an industry leader on the components side, Corning boasts an impressive diversity of technologies and products under development internally or acquired through strategic acquisitions. These include all-optical switches, polarizers, band splitters, WDM multiplexers, pump lasers, source lasers, photodetectors, modulators, and tunable filters.

Notably, Corning has become the world's largest supplier of erbium-doped fiber amplifier (EDFA) modules. EDFAs amplify, all-optically, WDM frequency bands simultaneously in a single fiber, making them crucial enablers of all long-distance fiber-optic networks. Corning was the first to capture an additional band of usable frequencies from 1,570 to 1,625 nanometers and the first to offer a hybrid Raman/EDFA whose backward-facing Raman pump lasers begin amplification before the signal reaches the amplifier itself. Launch power can thus be reduced and along with it troublesome signal distortions. Corning's new generation of variable-gain EDFAs will push customization by adjusting for disparate amplifier spacings among networks. Further on, the company may open up the 1,400-nanometer region for long-distance transmission with an all-optical amplifier probably based on thulium-doped fiber.

Moving optical component manufacturing out of the craft-guild era and into the industrial age represents one of the telecosm's greatest challenges. Corning commands as many patents in process technology as in the famous materials; its joint venture with Samsung to automate the manufacture of thin-film filters has already improved production yields by some 25 percent and reduced cycle times from two weeks to two days.

Along with JDS Uniphase, Corning can lead the way to cheap components for a ubiquitous fibersphere.

Corvis

In the telecosm, a great divide separates the companies performing the mostly empty stunt of multiplying the number of inaccessible bits per second on a backbone fiber from the companies multiplying the number of customers who can access the network. It separates the companies focused on big, costly bandwidth, such as Nortel and Lucent, from the companies focused on cornucopian connectivity, such as Corvis.

As Nortel advances its OC-768 technology over the coming years, Corvis's founder and optical guru, David Huber, will be enabling networks to eliminate SONET regenerator sites and electronic switching nodes on links of all distances, but chiefly in the lower billions of bits per second. Along with the sites go the infrastructure costs of the lasers and air conditioners and inventory and maintenance and cramped central office space.

On an average Los Angeles-to-Washington, D.C. link in an OC-192 opaque network, like Nortel's, bits will pass through seventeen SONET regenerator sites and five electronic switching sites for a total of forty-eight oeo (optical to electrical to optical) interfaces. In a Corvis transparent network, that same link requires only four optoelectronic conversions by eliminating five switching and fifteen regenerator nodes.

In addition to higher network build-out and data-transport costs, Nortel will also be saddled with inferior quality of service. Passing through electronic mazes adds delay (or latency) regardless of distance. In a Corvis network of seamless light paths, these delays vanish along with the oeo interfaces that cause them. In the course of its fifteen-week start-to-finish build-out of Broadwing's national all-optical backbone, the carrier suffered a fiber cut in its old network and asked Corvis to light a portion of the all-optical installation to make up for the cut. Broadwing's customer, a Forbes 100 software firm, called the networker immediately and told Broadwing: "Whatever you did, don't change it back!" so sharp was the improvement in network quality.

The implications of the all-optical network go beyond the elimination of SONET. Corvis has demonstrated in its network models what I have been preaching for years: that the most cost-effective, most efficient optical networks waste bandwidth. In Corvis language, when network traffic exceeds the capacity of merely one lambda on a data link, the network scales best by provisioning new lambdas for new links rather than conserving bandwidth by aggregating bits and switching old lambdas. Between major network hubs, Corvis establishes semipermanent express lambda lanes, which seldom need switching, as the most cost-effective solution to network management. When wires are cheaper than switches, there is scant need to switch at all.

Hence, it will be increasingly more cost-effective to run an all-optical network at 2.5 gigabits per second than a hybrid network at 40 gigabits per second. With many partially filled lambdas, the most "efficient" photonic networks will operate at a fraction of capacity. "Corvis" carriers will get rich on bandwidth "glut."

Because of the resistance of some carriers to all-optical networks, trust will be built up only when the economics become so compelling that they have no choice—as when an insurgent carrier forces the issue. Broadwing has taken up the challenge.

In the spring of 2001, Corvis completed Broadwing's nationwide all-optical backbone—18,500 WDM miles. Hardened with Corvis's small 6x6 all-optical switches and enabled by ascendant Raman amplification, an area where Corvis likely leads, Broadwing can please four million Napster downloaders per minute or supply 50,000 rabid readers with an eBook each during the same sixty seconds. With the potential of thousands of lambdas per fiber, thanks to Corvis, Broadwing and broadband could become synonymous.

Lighting a Corvis lambda is not a multimillion-dollar SONET project; the carrier can simply turn lambdas on as needed by adding cards to each end of the link. By using gain-flattening algorithms Corvis avoids sending technicians into the field to tweak each amplifier to accommodate the new lambda(s). And although it takes more effort to light a whole new fiber since new amplifiers and switches and dispersion compensators need to be deployed, Corvis was able to install Broadwing's entire national network in only fifteen weeks.

To light a new lambda (not a new *fiber*), SONET carriers might make use of Lear jets (actually done by one company) to do a coast-to-coast drop of circuit packs at every network node for the electronic switches and SONET regenerators required for each new lambda. And with the increased capacity, latency actually becomes worse.

Nortel counters that most network failures are in optics. But millions of light streams are more reliable than a few hundred. Today, Broadwing's all-optical Corvis-built network can scale to 160 channels, and Corvis hints at 700 channels available by year's end for anyone who can use it. In a broadband WDM domain, bandwidth scales in lambdas instead of bit rates—and as Corvis has shown us, it scales far more cost-effectively and elegantly than TDM.

JDS Uniphase

Built on the vision of its chairman, Kevin Kalkhoven, JDS Uniphase (JDSU), which George Gilder referred to as the "Intel of Telecosm," designs, develops, manufactures, and markets telecommunications equipment, including semiconductor lasers, high-speed external modulators, and transmitters for fiber-optic networks, the backbone of the bandwidth explosion. JDS Uniphase is the result of the merger between Uniphase and JDS Fitel, another juggernaut of the optical network component industry. Uniphase's laser division produces laser subsystems for a broad range of applications, including biotechnology, industrial process control and measurement, graphics and printing, and semiconductor equipment. The company has been sharpening its focus by divesting nonstrategic business units. The company's vision, to be the largest independent supplier and one-stop-shop for optical networking equipment, is now reality. The merger has insured this since with Uniphase focusing primarily on active components, and JDS Fitel the leader in passive components, there is virtually no overlap in the two companies' product lines.

Uniphase had developed a strong global presence with its acquisition of Philips Optoelectronics located in The Netherlands, which produces high-performance semiconductor lasers, photodiodes, and components for telecommunications, CATV, multimedia, and printing markets, including 1,550 nm, 1,310 nm, and 1,480 nm lasers, electro-absorption (EA) modulators, semiconductor optical amplifiers and receivers. Uniphase, with the opening of its European semiconductor laser fab in Zurich, moved a step closer to industry domination. The new plant will produce lasers to power the erbium-doped fiber amplifiers (EDFAs) which amplify optical signals as they pass through the fiber. EDFAs are created when standard glass fiber is treated, or "doped" with erbium, a rare-earth element, and activated with a pump laser.

Fiber Bragg Gratings are a critical element in the WDM system. Uniphase has created a new Network Components division (formerly its Indx acquisition) to develop Bragg Gratings, as well as channel add-drop modules and dispersion compensation modules. The company built a new production facility in Australia to produce the gratings last year. There is strong demand for these gratings, and orders already total more than the first year's planned production of this facility. Wavelocker, a new product, will enable the increasing numbers of channels in WDM systems. As the available space in a fiber becomes increasingly congested, Wavelocker allows source lasers to maintain wavelength stability by adjusting for temperature variation, and prevents interference among fiber channels.

JDS Fitel had also been strengthening its position through acquisitions and strategic alliances, including the photonics division of Akzo Nobel, which is a leading supplier of waveguide technology for optical switching. These planar waveguide-based switches use specialized optical polymers for low loss and increased flexibility of application. The switches will be useful for applications as add-drop multiplexing, optical cross connects, and protection switching. To maximize this technology, JDS Fitel has entered into a strategic alliance with AlliedSignal, a pioneer in the development of optical polymers for telecommunications applications. By bringing together AlliedSignal's material and process expertise with the newly acquired waveguide technology, JDS Fitel has enhanced its optical switching portfolio. In December 1998, the company announced a contractual joint venture with Corning Incorporated to develop and manufacture optical isolators. JDS Fitel designs and manufactures the isolators, while Corning supplies certain materials and other needs for the project.

The cable TV (CATV) industry presents a new growth opportunity for JDS Uniphase. The cable plant of the CATV industry is currently being upgraded to an HFC system, which will permit the use of WDM. TCI is the leader in the installation of cable WDM systems, and others are following given the great advantage in capacity. Uniphase has launched WDM components tailored for CATV systems, and is poised to supply this emerging market.

An emerging optical market is the burgeoning submarine (undersea) optical networking infrastructure. Equity Research projects the needs for pump lasers for these undersea networks alone could reach $300 million over the next four years. Uniphase makes the only qualified (Bellcore standard) submarine pump lasers currently on the market. The total fiber deployed in undersea fiber-optic networks could go as high as 1,700,000 fiber miles during this explosive build-out phase with twenty networks under way and over twenty more in various stages of development.

ONI

In the smaller metro and regional networks, where WDM has long been proclaimed too expensive, ONI continues to defy conventional wisdom. Building a captive customer base (now an impressive twenty-four, up from seven a year ago) primed for future upgrades, ONI's relentless CTO Rohit Sharma continues to succeed against competitors Nortel and Ciena in this rapidly growing space. ONI's recent 31 percent quarter-to-quarter revenue increase trounced Wall Street estimates and proved that connectivity matters much more than content since the former is what customers are willing to pay for. To get a foot in the carrier door, CEO Hugh Martin and CTO Rohit Sharma currently straddle the great divide, obliging telco customers with SONET ports until its demise. But their remotely reconfigurable WDM system currently scales to a connectivity-rich 160 channels per fiber—outer space compared to what metro networks use today.

Sharma takes network intelligence out of big Cisco core routers and switches and places it in servers on the edge where, as he explains, most of the network intelligence already resides. As a result, grooming—the capability to dynamically mix, match, and stack lower bit-rate data on WDM channels, inherently a bandwidth-conserving activity—also recedes from the core where bandwidth is abundant (and where Ciena's CoreDirector switch and Nortel's SONET now perform the function) toward the network edge where it is needed. There is a trade-off between packing lambdas efficiently and wasting them for connectivity. With bandwidth a replacement for grooming in network cores, Ciena's CoreDirector electronic grooming switches, for example, do you no good. To assure that every lambda is filled to the brim entails complicated and costly network management and statistical multiplexing equipment.

ONI technology makes lambdas on the edge more flexible, easier to deploy, and cheaper while reducing power consumption and conserving precious central-office space. Its 160-channel WDM system replete with dynamic add-drop and real-time lambda management now includes, with Corning's help, EDFA gain that is tunable within milliseconds, adapting to channel fluctuations on the fly. ONI offers end-to-end lambda management and provisioning from network cores to enterprise campuses and high-rises.

The forever forward looking Sharma—whose San Jose–based company, which plows a quarter of revenues back into research, recently opened a new optical research laboratory outside the U.S. in Montreal—plans to extend ONI's reach into Ethernet networks in 2002, having already added Ethernet interfaces to his WDM equipment and having acquired gigabit Ethernet multiplexing technology earlier in this year through his acquisition of Finisar's Opticity line.

Qualcomm

Qualcomm Incorporated (QCOM) is a leading supplier of digital wireless communications products and technologies. The company develops, manufactures, markets, licenses, and operates communications systems for a range of markets and applications.

Following the Gilder axiom of "wide and weak," code division multiple access (CDMA), pioneered by Qualcomm, is the winner, and has been designated to become the next generation of worldwide wireless transmission technologies. Over the course of the next three to six years, other competing forms of mobile wireless technology will migrate to the new higher bandwidth CDMA platform.

Currently CDMA is not the most widely used form of wireless, but it is growing rapidly because of its quality and efficiency. CDMA is a form of "spread spectrum" wireless and uses advanced digital encoding techniques. It allows a large number of users to share the same radio frequencies, and increases capacity some ten times over today's analog cellular systems.

CDMA encodes a voice or data signal so that it appears different from all the other transmitted signals. All signals occupy the same spectrum, which greatly increases capacity, but only the receiver attuned to a particular signal will receive it. This has been compared to a roomful of people all talking at the same time, but each speaking in a different language. The listener will be able to hear the person who speaks his language, and be able to "tune out" the other speakers.

Qualcomm also produces Eudora, a leading e-mail program. With an installed base of over 18 million users, Eudora software is one of the world's leading Internet e-mail packages. A new service recently launched is the Eudora e-mail by Phone, which uses text-to-speech technology to convert e-mail messages into digitized speech over a telephone. If you are unable to get to your e-mail through the usual channels, you can dial up and receive your messages over the phone.

Qualcomm's pdQ phone is an example of the convergence of digital devices. It functions as a CDMA or analog phone and as a Palm (Pilot) personal organizer. It has a built-in wireless modem (up to 33.6 kbps) and runs Palm 3.0 software including date book, address book, e-mail features, and web browser.

Following a March 1999 agreement with Ericsson, including the sale of its infrastructure business, Qualcomm can now concentrate on its core businesses of CDMA chipsets and handsets. The company is positioned to dominate a rapidly growing market in these chips, selling to not only second generation wireless vendors but soon to third generation CDMA wireless customers.

By all important technical measures, such as dropped call performance, call capacity and voice quality, CDMA is exceeding expectations. CDMA has experienced a more rapid deployment and growth in subscriber numbers than any other wireless technology, and has been deployed in over twenty-five countries. There are still some issues to resolve as to which CDMA technology will be chosen as the third generation standard. A single worldwide CDMA standard, which most would prefer, may not happen. Qualcomm owns the important intellectual property rights to CDMA, and regardless of the outcome of these individual skirmishes, is poised to lead the wireless world into the next century.

Texas Instruments

Texas Instruments (TXN) has embraced the digital future. The company envisions a time in the near future when every connection in the digital society will be made through a digital signal processor (DSP). DSPs are exceptionally fast and versatile semiconductor devices which are the engines driving the new digital society. Texas Instruments is the market leader in DSPs. The company has focused on DSPs and mixed signal devices, divesting itself of nonstrategic business units in an effort to build on its lead in these important technologies. In October 1998, the company sold its computer memory division to Micron Technology, a divestment which frees resources for more focused development. More than a dozen operations have been sold in the past few years, and many others have been acquired.

TI's strategy has merit. DSPs have a broad range of applications. These chips are found in personal computers, cellphones, networking equipment, consumer electronics, office equipment, and automotive components among other applications. The growth in the DSP market has been rapid, expanding globally at more than 30 percent per year since 1988. DSPs and mixed-signal devices are expected to outperform the general semiconductor market in the coming years.

TI is creating complete systems on a single chip. TI revolutionized this single chip market in 1996 by introducing the industry's first complete Ethernet switch on a chip. Since then others, such as National Semiconductor, LSI Logic, Analog Devices, and Atmel, have joined the effort driven by the constant push to create smaller, more power-efficient devices. Following the Gilder Paradigm corollary of "the less the space, the more the room," these chips have become smaller, contain a greater density of transistors, and, paradoxically, run cooler, faster, and consume less power. Texas Instruments has achieved a high level of systems integration in DSPs. The company has a wide range of component cores needed to develop and deliver various systems on a chip.

The company sells to a broad range of customers—currently more than 30,000 worldwide—most of which buy standard digital signal processing products. TI can also provide custom chips, working closely with customers to create DSPs and single chip systems for customer specific applications.

An attractive business for TI is digital subscriber line (*x*DSL), a telecommunication technology that allows more data to be sent over standard copper phone lines. DSL digital signal processors are used on both ends of the telephone wire, a potential market of 1.5 billion DSPs worldwide. The company is well-positioned to go after this emerging market because of its acquisition of Amati, whose DMT technology was selected as the standard for both American and European ADSL (asynchronous DSL) equipment. TI has new DSL chipset technology that is fully programmable, providing advantages such as scalabilty to handle future growth, and the flexibility to add new features and enable new applications.

Texas Instruments and NEC Corporation announced the production of engineering samples of very high bit-rate digital subscriber line (VDSL) modules. This DSP-based technology for modems, with a downstream speed of up to 52 Mbps, targets the delivering of real-time video and real-life, high-quality teleconferencing. TI has had declining revenues over the past several years, but with rising DSP revenues, has made a comeback with its refocused strategy. It currently commands nearly 50 percent of the market in DSPs.

As these DSPs become adapted to increasing numbers of applications, the company is poised to take advantage of its leadership in this rapidly growing chip market.

• • •

These nine companies see the future clearly, and it is bandwidth. Throwing aside the scarcities of the past, and enabling the digital age, they will give us new access to the world and to each other.

Appendix C

The Telecosm Glossary: An Opinionated Lexicon

A

ADC: See ***Analog-to-Digital Converter***

Add/Drop Multiplexing: A process by which traffic is inserted (added) or removed (dropped) from a network at some point in an end-to-end transmission. Very expensive using electronics that must process the entire data stream, potentially very cheap in optics that must only filter out a wavelength.

ADSL: See ***Asymmetric Digital Subscriber Line***

Advanced Intelligent Network (AIN): AIN (Antiquated Industry Nostrum) is a telephone network architecture that enables the public switched telephone network (PSTN) to offer a variety of services better performed by PCs and other devices on the edges of the network. Some services made possible by AIN include ***virtual private networks*** (VPNs) and caller ID.

Advanced Mobile Phone Service (AMPS): Yet another reason (see above) never to name your technology "advanced." AMPS is the prevailing U.S. standard for analog cellular phone communications, now very retarded. Based on ***Frequency Division Multiplexing*** (FDM), AMPS is some fifteen times less efficient than new ***CDMA*** digital systems.

Advanced Research Projects Agency Network (ARPANET): The ***Wide***

Area Network (WAN) that was established by the Department of Defense's Advanced Research Project Agency in the 1960s. Running at 50 kilobits per second, it is the network on which the ***Internet*** was originally based, but its relationship to the current Internet resembles the relationship of a military "supercomputer" of the 1960s to a PC of today.

Agent: A mediator between the ***client and server*** on a network that performs some user-defined task, especially in ***Internet applications. Search engines*** or ***spiders*** are sometimes referred to as agents, as well. Conceived before the popularization of the Internet, an "agent" is merely a marketing name for a program that operates outside of its host. Soon agents will be ubiquitous. With new technologies combining ***Java*** and ***Jini,*** for example, every device will contain its own agent.

AIN: See ***Advanced Intelligent Network***

ALOHANet: A simple and elegant network developed by Norman Abramson of the University of Hawaii that is based on *Time Division Multiple Access* (TDMA). Sharing ***bandwidth*** by issuing acknowledgments if packets are received and retransmitting lost packets, ALOHANet is used for satellite and radio links, and inspired Robert Metcalfe to invent the ***Ethernet.*** As Metcalfe pointed out in his book *Packet Communications,* Ethernet was "merely an ALOHANet ripoff" . . . except that Ethernet lacked (i.e., did not need) centralized control, used a carrier sense procedure, a collision detection system, and a random increasing backoff algorithm—to radically reduce collisions—and employed a single transmission channel with two address packets (rather than two channels with a single address packets), and "runs at millions of bits per second over cable within buildings, while ALOHANet ran at thousands of bits per second over radio among the Hawaiian Islands."

Amplification: The process of increasing the strength of a current, voltage, or various input signals (e.g., broadcast TV) in order to increase its strength and quality. *Contrast with* ***Attenuation.***

Amplitude: The height of a wave, measured from the baseline to the peak of the waveform. Amplitude determines the signal strength or power, which is calculated as the square of the amplitude (i.e., a four-foot wave is sixteen times more powerful than a one-foot wave . . . check it out on the beach.)

AMPS: See ***Advanced Mobile Phone Service***

Analog: A signal, such as voltage or sound, that represents data by varying—or ***modulating—amplitude*** or ***frequency*** in imitation or simulation of the image or sound to be transmitted. The human voice is analog, as are traditional cellular networks. Contrast with ***Digital*** which represents a phenomenon not by imitating it but by translating it to numbers, which in turn are sent in the form of bits. Because analog signals use all their features to convey information, analog is theoretically far more efficient than digital, which reduces a complex signal to a series of ones and zeroes. Nullifying this analog advantage in communications is a required signal-to-noise ratio at least a thousand times higher than digital.

Analog Display Services Interface (ADSI): A ***protocol*** developed by Bellcore that allows complex circuit switched networks to provide incoming Caller ID to a user occupied on another call—a function that is a "no brainer" for PCs performing ***IP Telephony.***

Analog-to-Digital Converter (ADC): A device that converts ***analog*** signals to ***binary digits*** (bits) that can be read by computers. The resolution or accuracy of the device is measured in bits—i.e., a 12-bit ADC can transform an analog input into a number as high as 2 to the 12th or 4,096. *Contrast with* ***Digital-to-Analog Converter.***

API: See ***Application Programming Interface***

Applet: A software application or component that is small, and designed to execute a specific task, often as part of an ***HTML*** page. A ***programming language*** that is commonly used to write applets is ***Java.*** Eventually, nearly all programs will consist of applets.

Application: A sequence of instructions, usually written in a higher-level language, that normally cannot be executed on a computer until it is compiled into a computer's own unique machine language—the stream of ones and zeroes that the computer can read. Some commonly used applications include word processors, browsers, and spreadsheets. *Also referred to as* ***software*** *or a program.*

Application Programming Interface (API): A set of routine ***software*** and ***hardware*** procedures implemented by the ***operating system*** to increase software performance, and consolidate functions that are common to several ***applications***—so as to eliminate rewriting the same command for each application.

Application-Specific Integrated Circuit (ASIC): A ***chip*** that contains a collection of initially unconnected logic ***gates*** that don't have to be linked to perform any specific function until late in the manufacturing process. Differentiated from general purpose ***microprocessors*** that execute a variety of ***software*** programs. *Also called* ***gate array.***

ARPANET: See ***Advanced Research Projects Agency Network***

ASIC: See ***Application-Specific Integrated Circuit***

Asymmetric: Used in reference to communications technologies in which the ***upstream*** and ***downstream bandwidths*** differ—e.g., ***Asymmetric Digital Subscriber Line.***

Asymmetric Digital Subscriber Line (ADSL): An ***asymmetric*** variant of ***Digital Subscriber Line*** (DSL) technology that can provide up to 9***Mbps downstream*** and 800***Kbps upstream***.

Asynchronous: Operating without a clock or timing device of any kind. As the move to single-***chip*** systems accelerates to the point that clock pulses can no longer reach across an entire chip in the time a ***transistor*** can switch, most ***microprocessor*** architectures will become at least partly asynchronous. See application in ***Asynchronous Transmission.***

Asynchronous Transfer Mode (ATM): A method of data, voice, video, and ***frame relay*** transmission. An ATM network provides a constant stream of addressed short "packets" called cells, each containing 53 ***bytes***, chosen in the mid-1980s as the largest ***packet*** size then with ***latencies*** acceptable for voice. Transferring multiple data streams and formats at the same time, an ATM network can actually simulate much more slowly the performance of a much cheaper and simpler optical ***wavelength division multiplexing system***.

Asynchronous Transmission: A method of transmitting data in which the data is divided into ***packets***. In this method, a "start" or "stop" component is added to each packet. Each packet is then sent independently to the receiving party. Synchronous transmissions use a clock pulse in order to avoid the overhead of start and stop bits.

ATM: See ***Asynchronous Transfer Mode***

Attenuation: The loss of transmitted signal strength. This loss can be a result of interference, or the inverse square law weakening of the signal as it travels further from its source, or, devices such as volume controls can induce attenuation deliberately. *Contrast with* ***Amplification.***

B

b: See ***Binary Digit***

B: See ***Byte***

Backbone: The portion of a communications network that connects smaller, local networks to each other. The backbone is ordinarily assumed to carry the majority of the traffic within a network, but in fact the law of locality indicates that at least 80 percent of the traffic will remain in an invertebrate ***Local Area Network.***

Bandwidth: The amount of data that can pass through a transmission medium. For example, the bandwidth of a ***T-1*** line would be 1.544***Mbps.*** Presently, most homes have 56.6***Kbps*** bandwidth available to them over the copper telephone local loop (i.e., copper cage). In telephone or cable TV systems, the bandwidth available is dependent on the connection—e.g., ***twisted-pair,*** coax cable, or ***fiber optics***—at the client's end, and the type of equipment at the central office or ***head end.*** But in ***dumb networks,*** the bandwidth is dependent on the transmission medium alone.

Baseband:

1. The base ***frequency*** of a signal before it is modulated onto a higher ***carrier*** frequency (i.e., ***microwave***) for transmission.

2. A form of transmission by which the medium (e.g., ***fiber-optic*** cable, wireless, or copper wires) transports a single signal at a time. This usually means that the digital signal is sent directly through the medium, rather than being divided into packets, as in ***broadband.*** However, baseband communications are utilized in ***Ethernet*** transmissions, which is a packetized system.

Base Station: The stationary location from which a mobile transmitter/receiver sends and receives signals. In ***cellular*** services, the base station is referred to as a ***cell*** site.

Basic Input/Output System (BIOS): The ***software*** on a computer that directs the system ***hardware*** (i.e., keyboard, floppy drives, and hard drive), and often runs tests on the hardware, at startup. The BIOS is stored in ROM (***Read-Only Memory***).

Basic Trading Area (BTA): A BTA is a geographical region of the U.S., as defined by the ***Federal Communications Commission's*** (FCC) revisions of the Rand McNally 1992 Commercial Atlas and Marketing Guide, based on certain economic and industrial criteria, in order to allocate areas of service. The FCC licenses each BTA to a service provider. Each BTA is made up of at least one county. There are presently 493 BTAs in the U.S. *Contrast with **MTA.***

Baud: The number of signal changes, or events, that take place in one second. This term is used in measuring ***modem*** speeds. It differs from ***bps*** in that through encoding processes, more than one bit can be represented in one event.

BBS: See ***Bulletin Board Service***

Bellcore: Bell Communications Research. A company formed by the ***RBOCs*** to develop technologies and standards to be utilized in their communications systems. Bellcore was sold to Science Applications International Corporation (SAIC) in 1997.

Bell Operating Company (BOC): See ***Regional Bell Operating Company***

Binary Digit (Bit): The smallest measure of information in digital systems. Bits are represented by either (1) or (0), and are typically represented physically by turning an electrical current on or off. A bit is of little value by itself, but grouped together bits are used to transfer letters, numbers, and other forms of data. Eight bits grouped together make a ***byte.***

BIOS: See ***Basic Input/Output System***

Bit (b) (*contrast with* B for Byte): See ***Binary Digit***

Bit Rate: The speed at which bits are transmitted in a communications medium—e.g., ***fiber optics,*** air (wireless), or ***twisted-pair.*** Bit rate is measured in ***bps.***

Bit Stream: A continuous flow of ***binary digits*** (bits), through some form of communications medium—e.g., ***fiber optics,*** air (wireless), or ***twisted-pair.***

BOC: Bell Operating Company. See ***RBOC***

Boolean Logic: A form of Algebra that was developed by the nineteenth-century mathematician George Boole. Through Boole's method, variables are not represented by numbers, but rather by "***Boolean operators.***" The result of a Boolean expression can only be one of two solutions—true or false. Because the absence (false) or presence (true) of electrical voltage easily represents these two results, Boolean logic is the basis for computer ***processors.***

Boolean logic is also an efficient method of conducting database searches, e.g., a search with the expression "Boolean AND Logic" would return only the records in a database in which both words are present.

Boolean Operators: AND, OR, NOR, NAND, XOR, or NOT. These terms are used in ***Boolean logic*** to represent variables.

Bot: The slang term for "Internet Robot." A program that searches the ***World Wide Web*** (WWW) for data, and indexes it in a database for later searches. Any program that performs repetitious or time-consuming tasks can also be considered a bot. Also referred to as a ***spider.***

bps: Bits per second. See ***Bit Rate***

Bragg Grating: A process in which multiple lines are etched into a ***fiber-optic cable*** to form a type of filter or reflector. These gratings act as prisms by diffracting light within the fiber. By using these gratings, an optical network adds the ability to select, divert, or focus light from within the fiber, negating the need for additional equipment to perform these functions.

Bridge: A device used to connect networks.

Broadband: Transmission mediums providing ***bandwidth*** equal to or greater than 1.544***Mbps.*** *Contrast with* ***narrow-band.***

Browser: An application used primarily to view ***HTML*** documents, primarily on the ***Internet.*** However, with the introduction of ***Java,*** ActiveX, and other ***plug-ins,*** the browser has evolved to provide support for much more than HTML. Popular browsers include Microsoft's Internet Explorer and Netscape's Navigator.

BTA: See ***Basic Trading Area***

Bulletin Board Service (BBS): A computer configured to share information, usually on a specialized topic. The BBS is established on a computer that is equipped with a ***modem*** to establish connections with other computers. Callers can post messages to other viewers of the BBS, and can often download various files pertinent to the BBS's topic. Functioning as a rudimentary online service, many BBSs later evolved to offer ***Internet*** access to their members, and have increasingly migrated to web-based services.

Bursty: The transmission of data in bursts, rather than the continuous flow of a ***bit stream.***

Bus: The transmission paths that allow computer components to contact each other. In addition to the ***hardware*** already connected to the bus—e.g., hard drive, ***microprocessor, I/O*** connections—most computers have several expansion slots that allow for other components to be connected to the bus.

Bus Speed: The communication speed between a computer's ***Central Processing Unit*** (CPU) and other system components.

Byte (B) (*contrast with* b for bit): A unit of measure that is made up of eight ***bits.*** A byte is capable of representing a single character such as a letter or number. In some instances—e.g., ***modem*** communications and nine-track tape storage—a byte is made up of nine bits, the ninth being a parity bit, or a bit used for checking accuracy in transmission.

C

C: A computer language supported strongly by UNIX, ***MS-DOS,*** and a wide variety of other ***operating systems.*** Dennis Ritchie developed the language at Bell Laboratories in the 1970s.

C++: An object-oriented programming language that is based on ***C,*** and has been adopted by many programmers. Bjarne Stroustrup developed C++ in the early 1980s at Bell Laboratories.

Cable Miles: Also known as sheath miles. The measurement, in miles, of ***fiber-optic cable*** that is deployed. *Contrast with* ***fiber mile*** *and* ***route mile.***

Cable Modem: A ***modem*** that takes advantage of the ***bandwidth*** available from a combination of ***coaxial*** and ***fiber-optic*** cables. Cable modems provide up to 40Mbps connections.

Cable Television (CATV): Also Community Antenna Television. A means of transmission by which broadcast signals are first received by an antenna at the service provider's ***head end,*** and then divided into small ***frequency*** "slots," or ***channels,*** that are sent to a subscriber's home via a ***coaxial cable.***

Cache:

1. Very fast ***memory,*** often ***SRAM,*** which is used by a computer to speed its operation. The cache retains the most frequently used information, so that the ***CPU*** does not have to access its slower memory (i.e., ***DRAM***) repetitively.

2. The storage of HTML pages or other network data locally—either on the end user's computer or an ISP or corporate server—to speed access by the end user.

CAP: See ***Competitive Access Provider***

CATV: See ***Cable Television***

CCD: See ***Charge-Coupled Device***

CDMA: See ***Code Division Multiple Access***

CDPD: See ***Cellular Digital Packet Data***

CD-ROM: See ***Compact Disc Read-Only Memory***

CDSL: See ***Consumer Digital Subscriber Line***

Cell: The geographic area that is serviced by a single radio transmitter/receiver in a wireless telephone system, or ***cellular*** system. A cellular system is made up of multiple cells, designed to provide seamless connections as a user travels from cell to cell. Cells vary in size based on what type of interference (e.g., terrain, existing radio frequencies, and buildings) is present in the cell.

Cellular: A form of radio transmission by which geographic areas are split into ***cells,*** that are serviced by individual transmitter/receivers, or ***base stations.*** These base stations have the ability to hand calls to each other, meaning that a user can maintain continuous connections rather than going out of range, as in a traditional radio transmission. Cellular technology is used to provide wireless telephone service.

Cellular Digital Packet Data (CDPD): A method of wireless data transmission developed to operate over existing voice networks. CDPD systems break data into packets that are transmitted over unused voice channels.

Cellular Telephone Industry Association (CTIA): A national organization that represents cellular and wireless communications companies.

Central Office (CO): A switching station that functions as an intermediary between the subscriber and the rest of the telephone network.

Central Processing Unit (CPU): The component of a computer that receives, interprets, and executes logic functions. The CPU is essentially the "brain" of the computer—telling the rest of the computer components what to do, and when.

CERN: See ***Conséil European pour la Recherche Nuclèaire***

CGI: See ***Common Gateway Interface***

Channel: A path through which communications are transmitted. A channel is used for transmission from one point to another, as in a standard telephone connection, or for a transmission from one point to multiple points, as in radio or television broadcasts.

Charge-Coupled Device (CCD): A device used as the image sensor in many scanners, as well as both video and digital cameras. Less expensive ***CMOS*** sensors threaten to replace CCD technology.

Chip: An ***integrated circuit*** that has been "packaged" in insulating plastic. The plastic casing traditionally has protruding metal pins that are used to connect with other chips to perform specific functions. Examples include ***microprocessors*** and ***DRAMs.***

Circuit Switching: A method of communications commonly used by telephone companies. In circuit switching, a connection is established between two parties that remains open until one party disconnects.

CISC: See ***Complex Instruction Set Computing***

CLEC: See ***Competitive Local Exchange Carrier***

Client/Server: A form of networking architecture by which a single computer—the server—performs company-wide tasks such as data storage, network administration, and network security. Meanwhile, multiple computer workstations—the clients—run the applications and tasks specific to the individual user. This system takes advantage of the computing power provided by both the server and the client.

CMOS: See ***Complementary Metal Oxide Semiconductor***

CO: See ***Central Office*** (or ***Head End***)

Coaxial Cable: A cable used in ***CATV*** systems and computer networks, among other applications. Often referred to as "coax," this cable is made of a center conductive core that is wrapped in an insulator and then a second conductive layer. These three layers are then typically wrapped in an outer insulating layer. Coax can carry a much greater amount of data than traditional copper wires.

COBOL: See ***Common Business Oriented Language***

Code: A set of instructions, as defined by a computer programmer, which collectively define a software ***application.***

Codec: See ***Coder/Decoder*** or ***Compressor/Decompressor***

Code Division Multiple Access (CDMA): A form of ***spread spectrum*** communications in which a signal is coded with an identifier before transmission. The same spreading code that attenuates the signal so it does not interfere with other signals also weakens and spreads out any spikes of noise. The receiver then recognizes the identifier, and multiplies the signal by an inverse code that eliminates the spread and captures the signal, while once again spreading out and attenuating any noise. Through this method, signals do not have to be sent through specific ***channels,*** but use only the ***bandwidth*** necessary for that transmission. CDMA is a technology that was developed primarily by Qualcomm. *Contrast with* ***Frequency Division Multiple Access*** *and* ***Time Division Multiple Access.***

Coder/Decoder (Codec): A combination coder and decoder that can process signals in both directions of transmission, usually in a single ***hardware*** device. *See also* ***Compressor/Decompressor.***

COM: See ***Component Object Model***

COM Port: The serial interface on computers that allows for devices such as modems, mice, and some printers to be connected. Each device that is connected to the computer is assigned a number. For instance, if a modem and mouse are both connected to COM ports on the computer, the modem would be assigned the COM-1 port, and the mouse would be assigned the COM-2 port, or vice versa.

Common Business Oriented Language (COBOL): A ***programming language*** that was developed between 1959 and 1961. COBOL is one of the oldest programming languages and is still widely used today.

Common Gateway Interface (CGI): A ***protocol*** by which a web ***server*** interacts with a predefined ***application,*** in order to complete tasks that would not be possible using simple ***HTML.*** Requests are made from the client's browser, and then relayed via the web server to the CGI application. The application then returns a response, via the web server, to the client's browser. By this method, the application can be run on any platform, or ***operating system*** (OS). CGI can be written in many different programming languages, including ***C, C++, Java,*** and ***Perl.***

Common Object Request Broker Architecture (CORBA): A technology developed by the Object Management Group for the sharing of information between ***applications*** and objects within applications. An open standard alternative to Microsoft's ***OLE,*** favored by Netscape, Sun, and others. CORBA is used by the object oriented ***programming language, Java.***

Communications Satellite Corporation (COMSAT): A corporation that was established in 1962 by the US Congress to exclusively provide international satellite communications to the U.S.

Compact Disc Read-Only Memory (CD-ROM): A computer storage technology that utilizes laser optics rather than the magnetic technologies utilized by other data storage devices, such as floppy disks. Compact discs are read-only, and can hold approximately 650***MB.***

Competitive Access Provider (CAP): An ***Internet*** and/or telephone service provider established to provide competitive rates and services. Also: Contact Access Point.

Competitive Local Exchange Carrier (CLEC): A local telephone services provider. CLECs have proliferated in the deregulated environment created by the Telecommunications Act of 1996, and are typically small startups that compete with ***ILECs.***

Complementary Metal Oxide Semiconductor (CMOS): A type of ***integrated circuit*** (IC) that is composed of two ***MOSFETs***—one N-type and one P-type—that are integrated on a single chip. A CMOS consumes very little power, yet performs at very high speeds.

Complex Instruction Set Computing (CISC): A ***microprocessor*** that utilizes a very complex set of instructions. These instructions allow the processor to perform many complicated tasks very quickly, but tend to take longer periods of time to process simpler functions than competing ***RISC*** processors.

Component Object Model (COM): A method, created by Microsoft, used to build ***application*** components that can be assembled to create entire programs or add functionality to existing Windows programs. COM provides the basis for ***OLE*** and is the basic technology used in Microsoft's browser, Explorer. Compare with CORBA (Component Request Broker Architecture), which is often used with ***Java.***

Compressor/Decompressor (Codec): A device used to compress and decompress audio or video data. A codec, in this sense, can be either ***hardware*** or ***software,*** or a combination of the two. *See also* ***Coder/Decoder.***

COMSAT: See ***Communications Satellite Corporation***

Consèil European pour la Recherche Nuclèaire (CERN): European Nuclear Research Counsel. The physics research center—located in Geneva, Switzerland—that was the birth place of the ***World Wide Web.***

Consumer Digital Subscriber Line (CDSL): A form of ***DSL*** developed by Rockwell, that is slower than ***ADSL,*** but does not require as complex equipment on the subscriber's end.

CORBA: See ***Common Object Request Broker Architecture***

CPU: See ***Central Processing Unit***

CTIA: See ***Cellular Telephone Industry Association***

Cycle Time: The amount of the time between one ***RAM*** access and the next.

D

DAC: See ***Digital-to-Analog Converter***

Dark Fiber:

1. ***Fiber-optic cable*** that has been installed, but is not presently connected to ***opto-electronic*** equipment.

2. Fiber that has been installed without opto-electronic equipment by one company, but is later fitted with the equipment by a second party that leases the fiber from the installing company.

Data Over Cable Service Integration Specification (DOCSIS): A set of standards for ***cable modems.*** Initially developed by the MCNS—Multimedia Cable Network System, a partnership between a number of large cable companies—it has subsequently come to be managed by CableLabs, the research and development labs of the cable industry. The standards were written for the purpose of cable modem and ***head end*** system equipment interoperability. In addition to providing standards for broadband data transfer (i.e., ***Internet*** access), the standard is now being refined to add ***IP Telephony*** and other protocols. Converging on DOCSIS long before the telephone industry could arrive at one standard for its competing ***digital subscriber line*** technology (DSL) helped give the cable TV industry its decisive lead in bringing broadband Internet access to homes.

Data Rate: The speed at which data can be transmitted through a specific communications medium (e.g., ***fiber-optic*** cable or ***coaxial cable***). The data rate is typically calculated in ***bits per second*** (bps).

dB: See ***Decibel***

DBS: See ***Direct Broadcast Satellite***

Decibel (dB): A method of measuring sound or electrical signals. A decibel is one tenth of a bel (a unit of measurement, used to measure the strength of a sound or signal, that is named after Alexander Graham Bell). It is the ratio of two power levels, such as two sound or electrical signals, multiplied by ten. The decibel equivalents for numbers in factors of ten can be readily calculated by translating to scientific notation and then adding a zero to the exponent. Thus, the decibel translation of a factor of 10 is 10dB; 100 is 20dB; 1,000 is

30dB; and one thousandth is minus 30dB, and so on. Other useful dB equivalents include: 2 is 3dB; 3 is 5dB; 30 is 15dB and 2000 is 33dB. The dB equivalent for any number can be calculated by finding its logarithm on a calculator and multiplying it by ten.

DECT: See ***Digital European Cordless Telecommunication***

Demodulation: The process by which modulated carrier waves, or waves that are carrying data, are converted into digital signals. *Contrast with* ***modulation.*** *Also, see* ***modem.***

Demultiplex: Also known as DEMUX. To separate two or more channels that were ***multiplexed*** by compatible equipment.

Dense Wavelength Division Multiplexing (DWDM): An unnecessary marketing cosmetic for WDM, originally used for any system that used more than two wavelengths. Now virtually all WDM is dense, rendering the D otiose. Resembles telco's jumbo groups, super jumbo groups, and ultra jumbo groups.

Digital: A signal that represents data in the form of ***binary digits*** or numbers. Computers use digital signals in order to process information. *Contrast with* ***analog.***

Digital Cellular: A variant of ***cellular*** that uses ***digital*** signals, rather than the traditional ***analog*** signals. The use of digital cellular provides a much more reliable connection by eliminating static and easing congestion at the ***cell*** sites.

Digital European Cordless Telecommunication (DECT): A pan-European standard for wireless services. DECT is based on ***TDMA*** technology, and is primarily used in such applications as wireless PBX and LAN systems. DECT utilizes the 1800–1900 MHz frequency range.

Digital Loop Carrier (DLC):

1. Equipment that is installed in a phone network that ***multiplexes,*** or combines, ***channels*** at the ***central office*** (CO), and then ***demultiplexes*** the channels at a remote termination (RT) point to be delivered over separate ***twisted-pairs*** to customers. This decreases the number of actual twisted-pair copper wires that are required between the CO and the RT.

2. With the advent of ***fiber-optic*** cable, phone companies have begun to build networks consisting of fiber and twisted-pair copper wires. The DLC has come to mean equipment that not only handles the multiplexing/demultiplexing functions, as stated in definition 1., but also handles the ***optoelectronic*** conversion that is necessary when transmitting through these systems.

Digital Signal Processor (DSP): A specialized ***semiconductor*** that is designed to process a stream of ***bits,*** or a ***digital*** signal, in ***real time.*** A DSP typically has much greater mathematical computational abilities than a standard

microprocessor. They were pioneered by Texas Instruments in 1982, and are now found in thousands of devices; from ***DVD*** players to antilock brakes. DSPs are especially important in communications equipment, including ***modems*** and wireless telephone systems.

Digital Subscriber Line (DSL): A technology that provides high-***bandwidth*** connections to end users over standard ***twisted-pair*** copper wires employed by telecommunications companies. The data rates of these connections drop rapidly—due to signal ***attenuation***—as distance between the ***central office*** and the end user's ***modem*** increases.

There are many variations of DSL technology that have been developed. Some provide asymmetric transmission speeds, and some provide symmetric transmission speeds. The variations collectively are referred to as *x*DSL technologies. *See **ADSL, CDSL, DSL Lite, HDSL, IDSL, RADSL, SDSL, UDSL,** or **VDSL** for specific information on each.*

Digital-to-Analog Converter (DAC): A device that converts ***binary digits,*** or bits, to analog signals. *Compare with* ***Analog-to-Digital Converter.***

Digital Versatile Disk (DVD): Also known as ***digital*** video disk. A device that is expected to replace the ***CD-ROM*** because of its drastically improved storage size over the CD. Like CD-ROM technology, DVD players utilize lasers to read the DVD, rather than magnetic technologies employed by floppy drives. A typically DVD can hold up to 4.7GB of data—enough room for one full-length 133 minute movie, using ***MPEG-2*** compression.

DIMM: See ***Dual In-Line Memory Module***

Direct Broadcast Satellite (DBS): A high-powered satellite system used to transmit ***digital*** video signals directly to subscribers' homes. Subscribers receive the signal via a very small—eighteen inches in the Hughes DirecTV system—antenna, or satellite dish. The dish is attached to a small set-top box, which is attached to a television for the purpose of choosing the channel to display. DBS systems, such as DirecPC, have also been developed to provide high-speed ***Internet*** connections.

Discrete Multitone (DMT): A technology that is utilized by such transmission methods as ***DSL*** and ***ISDN.*** DMT makes possible 6Mbps data-rates over a single ***twisted-pair*** copper wire. DMT employs ***DSPs*** to split the available channel into many channels. This allows users to place voice calls and maintain a ***modem*** connection on the same phone line simultaneously.

Disk Operating System (DOS): A program that resides on a computer's hard drive, and is loaded when the computer is turned on. DOS is used to organize the computer's interaction with the hard and floppy drives, as well as the keyboard, mouse, and various other system devices. The most common version of DOS is MS-DOS, or Microsoft DOS. *See* ***Operating System.***

Dispersion: Scattering.

1. In computer terms, scattering is the storage of data on different computers or devices within an interconnected network.

2. In communications, scattering occurs as a path of light travels down a ***fiber-optic cable.*** As distinct light paths travel further from their source they begin to overlap, dispersing their signals. Dispersion within fiber systems is remedied through utilizing such equipment as ***Erbium-Doped Fiber Amplifiers.***

Distortion: A change in a ***waveform,*** between its transmitter and receiver, which usually results in information loss or garbled reception. When a radio signal becomes distorted, the reception typically becomes hard to understand, or even unrecognizable.

DLC: See ***Digital Loop Carrier***

DLL: See ***Dynamic Link Library***

DMT: See ***Discrete Multitone***

DNS: See ***Domain Naming System***

DOCSIS: See ***Data Over Cable Service Integration Specification***

Domain: A domain is a portion of a ***domain name.*** There are actually several subdivisions that make up a domain: the Top Level Domain (TLD), the secondary domain, and the tertiary domain. The TLD is the suffix found after the last period in a domain name, and is comprised of either a generic TLD (e.g., com, .edu, .gov), or a country code (e.g., .us, .fr). The TLD represents either the nature of business (generic TLD), or the geographic location (country code) of an organization. The secondary domain consists of a name that is unique to the entity that registers it. The secondary domain is found in the middle of the domain name. The tertiary domain is the prefix of the domain name. The tertiary domain is typically "www," but is not limited to those three letters. A tertiary domain is not always necessary, and may be excluded from some domain names. *See* ***Domain Name.***

Domain Name: The compilation of the various types of ***domains,*** separated by periods. Domain names are compiled with the tertiary domain listed first, the secondary domain listed next, and the Top Level Domain listed last, e.g., www.gildertech.com. Domain names are representatives of ***IP addresses*** that identify individual computers on the ***Internet.*** A principal benefit of domain names is that they provide addresses that are easier to remember than the twelve numerical digits of an IP address. *See* ***Domain.***

Domain Naming System (DNS): The system by which ***domain names*** are translated into the actual ***IP addresses*** that are assigned to computers on the ***Internet,*** and vice versa. The DNS is a database that is placed on domain name ***servers*** throughout the Internet. When a domain name is entered into a browser, it is first routed through a domain name server that translates the domain name into a twelve-digit IP address. The network is then searched for the computer assigned to that IP address.

Dongle: A small hardware component, called a "key," which is attached to an ***I/O*** port of a computer. The dongle validates a particular piece of software, and is generally used to prevent unauthorized reproduction of high-end software. Because the software will not run without a dongle, an additional dongle must be purchased for each computer that wishes to run the software.

DOS: See ***Disk Operating System***

Downstream: The path, in communications, through which a service provider transmits data to the end-user. *Contrast with* ***upstream.***

DRAM: See ***Dynamic Random Access Memory***

DSL: See ***Digital Subscriber Line***

DSL Lite: A variant of ***Digital Subscriber Line*** (DSL) technology that differs from ***ADSL,*** in that it does not involve splitting the line on the subscriber's end. Instead, the line is split at the ***central office,*** reducing the ***bandwidth*** of the line. A DSL Lite connection is typically capable of data rates of approximately 1.544Mbps or the equivalent of a ***T-1*** line, while ADSL data rates reach up to 9Mbps.

DSP: See ***Digital Signal Processor***

Dual In-Line Memory Module (DIMM): A small ***printed circuit board*** (PCB), on which ***RAM chips*** are mounted. A DIMM is equipped with connections that allow it to be plugged into sockets on a computer's main board. Relative to ***SIMMs,*** DIMMs provide a superior method of adding ***memory*** to a computer.

Dumb Network: A communications network that, in contrast to the ***public switched telephone network*** (PSTN), relies on information provided by the sender to route the transmission. A network in which the intelligence is located on the periphery—i.e., ***Ethernet.***

Duplex: A means of two-way communication that is capable of simultaneous transmission and reception in both directions.

DVD: See ***Digital Versatile Disk***

DWDM: See ***Dense Wavelength Division Multiplexing***

Dynamic Link Library (DLL): A file, used in the Windows or OS/2 ***operating systems,*** that contains information usable by several ***applications.*** DLL files, when edited by programmers, update or upgrade several applications at the same time.

Dynamic Random Access Memory (DRAM): A variation of ***RAM*** that stores each bit of data, as an electrical charge, in a cell that is made up of a capacitor and a transistor. Because capacitors lose their charge rather quickly, the cell must be refreshed with a new electrical charge every few milliseconds. If a ***processor*** is attempting to access the DRAM during this refresh period, it must wait until the refresh is finished. Thus, DRAM is slower than some other types of RAM (e.g., ***SRAM***), yet it is commonly used because of its superior storage size. Also ***Dynamic Random Excess Memory.***

E

E: See ***Exa***

E-: Electronic, as in ***E-mail*** or ***E-Commerce.*** Used as a prefix to designate a network-based service or process.

E-Cash: See ***Electronic Cash***

EB: See ***Exabyte***

E-Commerce: See ***Electronic Commerce***

EDFA: See ***Erbium-Doped Fiber Amplifier***

EEPROM: See ***Electronically Erasable Programmable Read-Only Memory***

Electromagnetic Radiation: A wave that is created through the acceleration of an electrical charge, and possesses both electric and magnetic characteristics. These waves make up the ***electromagnetic spectrum.*** The environment of human life and evolution, believed by some to cause cancer—by others to cure cancer.

Electromagnetic Spectrum: The entire collection of ***electromagnetic radiation,*** including: cosmic-ray photons, gamma rays, X rays, ultraviolet, visible light, infrared, microwaves, radio waves, audio transmission frequencies, and powerline distribution pulses.

Electronic Cash (E-Cash): A term referring to money that is exchanged over the ***Internet*** or some other form of electronic information technology.

Electronic Commerce (E-Commerce): Any commercial purchase or exchange of goods or services that takes place via some form of electronic information technology (e.g., the ***Internet***).

Electronically Erasable Programmable Read-Only Memory (EEPROM): A ***nonvolatile memory chip*** that is programmed subsequent to the manufacturing process. By applying an electrical signal, an EEPROM's programming can be erased, and subsequently reprogrammed. Slower than a

DRAM, but faster than a disk. EEPROMs are now being widely displaced by ***Flash Memory.***

Electron: The negatively charged subatomic particle that is the fundamental component of electricity.

E-mail: Electronic mail. A letter that is sent electronically (e.g., via the ***Internet,*** or a ***LAN***).

EPROM: See ***Erasable Programmable Read-Only Memory***

Erasable Programmable Read-Only Memory (EPROM): A ***nonvolatile memory chip*** that is programmed subsequent to the manufacturing process. Removing an EPROM's protective cover and exposing the chip to ultraviolet light erases its programming. It can then be reprogrammed. Now being displaced by ***Flash Memory.***

Erbium-Doped Fiber Amplifier (EDFA): An amplifier used in ***fiber-optic*** networks. An EDFA is a loop of fiber that has had erbium added to it through a process called doping. The EDFA is spliced into the fiber network, and powered by a ***pump laser.***

Erlang:

1. A unit for measuring telephone usage that is equal to one phone line being used for one hour (i.e., 60 minutes = 3,600 seconds, the equivalent of one Erlang).

2. A ***programming language,*** originally developed by Ericsson for use in the telecommunications industry, which is platform-independent. Erlang is similar to ***Java,*** in that ***applications*** written in Erlang use a ***virtual machine*** to run, allowing programs to be written once for many different platforms.

Ethernet: The most commonly used set of standards for networking computers. Ethernet was originally developed at Xerox under Robert Metcalfe, and later became the basis for the ***IEEE*** 802.3 standard for ***Local Area Networks*** (LANs). The 802.3 standard offers support of up to 100Mbps. It prevailed because it was medium independent and relegated all intelligence to the terminals on the network.

Exa- (E): A prefix that denotes 10^{18} or 1,000,000,000,000,000,000. In computer terms, however, exa- is actually equal to 2^{60}, or 1,152,921,504,606,846,976, the power of 2 that is closest to one quintillion.

Exabyte (EB): An exabyte is literally equal to 2^{60} ***bytes,*** or 1,152,921,504,606,846,976 bytes, but is often calculated with a base of 10, making it equal to 10^{18} or one thousand quadrillion bytes. LAN traffic in 1998 was measured roughly in exabytes per week; Internet traffic in ***petabytes*** per week.

Extranet: An extension of a corporate ***intranet*** that is set up to offer limited access to the company's suppliers and/or customers, utilizing ***Internet protocols*** and technologies.

F

FCC: See ***Federal Communications Commission***

FDDI: See ***Fiber Distributed Data Interface***

FDM: See ***Frequency Division Multiplexing***

FDMA: See ***Frequency Division Multiple Access***

Federal Communications Commission (FCC): Administers ***Moron's law.*** A United States government agency that was formed by the Communications Act of 1934. The FCC is charged with the duties of overseeing and regulating radio, television, wireline, satellite, and cable communications. Five commissioners direct the FCC.

FET: See ***Field-Effect Transistor***

Fiber Distributed Data Interface (FDDI): An American National Standards Institute (ANSI) standard that defines a 100***Mbps fiber optic Local Area Network.*** FDDI is based on a ***token ring*** network topology.

Fiber Miles: The measurement, in miles, of the actual length of fiber strands deployed in a network. A typical ***fiber-optic cable*** is comprised of multiple fiber strands that are bound together. Therefore, a fiber mile would be the measurement of the cable in ***cable miles*** multiplied by the number of fiber strands encased within that cable. The significance of this measure has been drastically altered by the advent of ***wavelength division multiplexing*** (WDM) which enables transmission of hundreds of separate bitstreams of information down every fiber strand.

Fiber Optics: A communications medium, in which data are transmitted, using a laser, through a thin strand of transparent material, most commonly glass.

Fiber-Optic Cable: A cable, used in ***fiber optics,*** that is made up of multiple strands of a transparent medium, called fiber. The most commonly used medium is glass.

Fiber-to-the-Curb (FTTC): A communications network in which the ***fiber-optic cable*** extends past every home, while the connection to the home itself is either ***twisted-pair*** copper wire or ***coaxial cable.***

Fiber-to-the-Home (FTTH): A communications network in which ***fiber-optic cable*** extends throughout the entire system, and directly into the home, eliminating the low-***bandwidth, twisted-pair*** copper wires.

Field-Effect Transistor (FET): A type of ***transistor*** that uses an electrical field between its two gates to control signals. The FET is turned on or off by pulses of electrical current. FETs are especially useful for ***amplification*** of weak signals.

Field Programmable Gate Array (FPGA): A ***programmable logic device*** that contains a large number of ***gates.***

Field Programmable Interconnect Component (FPIC): A technology that interconnects ***FPGAs*** with other components, and controls them through ***software,*** in order to allow electronic equipment designers the ability to prototype their designs before manufacture.

File Transfer Protocol (FTP): The protocol used for transferring files between two computers over a network, typically ***TCP/IP*** networks, such as the ***Internet.*** FTP utilizes its own set of commands.

Firewall: A security system that is placed within a network in order to restrict traffic to specific areas. Firewalls protect corporate ***intranets*** from being accessed by computers outside the network, and/or restrict the intranet computers from accessing outside resources.

Flash Memory: ***Electronically Erasable Programmable Read Only Memory*** (EEPROM) that can be entirely programmed or erased at once.

Floating-Point Operation (FLOP): The calculation, or processing, of floating-point numbers, i.e., numbers that, in relation to one another, have differing decimal point placements.

Floating-Point Operations per Second (FLOPS): The number of ***floating-point operations*** that can be accomplished in one second by a ***Central Processing Unit*** (CPU).

FLOP: See ***Floating-Point Operation***

FLOPS: See ***Floating-Point Operations per Second***

FPGA: See ***Field Programmable Gate Array***

FPIC: See ***Field Programmable Interconnect Component***

Frame:

1. A packet used for transmitting data in a communications network. Data is divided into chunks that are inserted into a frame of a predetermined size.

2. A section of an ***HTML page*** that can be scrolled and/or printed independently from the rest of the page.

3. A single full-screen image in a video clip, which when viewed in sequence with accompanying images, provides the appearance of motion.

Frame Relay: A method of networking that utilizes ***frames*** to transmit data. Frame relay can transmit data at rates of up to 1.544***Mbps,*** or ***T-1*** rates.

Frequency: The number of ***waveforms*** or cycles completed in a second. Frequency is measured in ***hertz*** (Hz). 1***Hz*** is the frequency at which one complete cycle occurs per second.

Frequency Division Multiple Access (FDMA): A form of communications that divides a channel into frequency bands, and assigns each signal its own band. FDMA is used in many ***cellular*** and ***PCS*** systems. *Contrast with* ***Code Division Multiple Access*** and ***Time Division Multiple Access.***

Frequency Division Multiplexing (FDM): A form of transmission whereby a single ***channel*** can transport multiple signals simultaneously. The transmission channel is divided into subchannels. Then, each signal is assigned its own subchannel.

FTP: See ***File Transfer Protocol***

FTTC: See ***Fiber-to-the-Curb***

FTTH: See ***Fiber-to-the-Home***

G

G: See ***Giga-***

GaAs: See ***Gallium Arsenide***

Gallium Arsenide (GaAs): A ***semiconductor*** material used to produce various types of ***integrated circuits.*** GaAs was originally developed for ***chips*** used in military communications, and has since been developed commercially. *Contrast with* ***silicon germanium.***

Gate: Also called a logic gate. An electrically controlled ***switch*** that produces a signal that represents either a (0) or (1). Gates are the basic component of a digital circuit.

Gate Array: A ***chip*** comprising a number of ***gates*** that are left unassigned until late in the manufacturing process. A layer can then be added to the chip that connects the gates in the necessary pattern for the desired function.

Gateway: The point in a network that provides a connection to another network. A gateway provides both the physical connection, as well as the conversion necessary to communicate with the other network's ***protocols.*** An example of a gateway would be an ***IP telephony*** gateway, connecting the ***Internet*** and the ***Public Switched Telephone Network,*** thereby allowing calls between a computer on the Internet and a traditional telephone, or between two telephones using the Internet as an intermediary for a long-distance connection.

Gb: See ***Gigabit***

GB: See ***Gigabyte***

GDI: See ***Graphical Device Interface***

GEO: See ***Geosynchronous Satellite***

Geosynchronous Satellite (GEO): A satellite that moves at the same speed as the Earth's rotation, thereby maintaining a stationary position relative to the ground. GEOs are utilized in a variety of different communications applications, but deployed 22,282 miles above the equator they suffer from a 250-millisecond roundtrip delay inflicted by the speed of light. Thus they are poorly suited for two way data or interactive operations. Contrast ***Low Earth Orbit Satellite.***

GHz: See ***Gigahertz***

Giga- (G): A prefix that denotes 10^9 or 1,000,000,000. In computer terms, however, giga- is actually equal to 2^{30}, or 1,073,741,824, the power of 2 that is closest to one billion.

Gigabit (Gb): A gigabit is literally equal to 2^{30} ***bits,*** or 1,073,741,824 bits, but it is often calculated with a base of 10 (10^9), making it equal to one billion bits.

Gigabit Ethernet: A variant of ***Ethernet*** that provides specifications for network speeds of up to one ***gigabit*** per second. Gigabit Ethernet is also referred to as ***IEEE*** standard 802.3z.

Gigabyte (GB): A gigabyte is literally equal to 2^{30} ***bytes,*** or 1,073,741,824 bytes, but it is often calculated with a base of 10 (10^9), making it equal to one billion bytes.

Gigahertz (GHz): 10^9 ***hertz*** or one billion hertz.

Gigaswitch: A communications ***switch*** that is capable of switching a ***gigabit*** of data each second. Gigaswitches are used to route ***Internet*** traffic in ***NAPs*** and ***MAEs.***

Global Positioning System (GPS): A satellite system by which geographic location is determined—including latitude, longitude, altitude, and precise time—via signals transmitted between specialized GPS satellites and ground receivers. These receivers can be hand-held or fixed-installation devices.

GPS: See ***Global Positioning System***

Graphical Device Interface (GDI): The Microsoft Windows display system. The GDI interacts with the display device or the printer to provide it with the information about the image that is to be displayed or printed.

Graphical User Interface (GUI): A method of computing that utilizes icons, or graphical images, to represent commands. A ***mouse*** is typically used to select, move or manipulate these icons. Some examples of GUI are Microsoft's Windows ***operating system*** (OS), as well as Apple's Macintosh OS.

Groupe Speciale Mobile (GSM): Or Global System Mobile. A standard for ***digital cellular*** communications, based on ***TDMA,*** that is used throughout Europe and much of the United States. GSM competed with ***CDMA*** technology until early in 1998 when the European Telecom Standards Institute endorsed ***CDMA*** for the next generation of GSM phone service.

GSM: See ***Groupe Speciale Mobile***

GUI: See ***Graphical User Interface***

H

Hardware: The physical components—including the ***CPU,*** printer, ***modem, mouse,*** and monitor—that are involved in the processing, displaying and/or communicating of data. Confusing this apparently lucid definition is the presence of millions of lines of ***software*** code embedded in all these devices in the form of microcode, "firmware," and other kinds of computer instructions. This fact became famous as the year 2000 approached and it became clear that what we call hardware was also sometimes afflicted with the "millennium bug" based on two-digit date fields.

HBT: See ***Heterojunction Bipolar Transistor***

HDSL: See ***High Bit-rate Digital Subscriber Line***

HDTV: See ***High Definition Television***

Head End: The point at which the signal originates (either generated or received from another source, e.g., satellite) in a communications network, such as in a ***cable television*** system.

Hertz (Hz): The unit used to measure ***frequency.*** One hertz is equal to one cycle per second.

Heterojunction Bipolar Transistor (HBT): A type of ***transistor*** composed of several different layers of ***semiconductor*** material. HBTs are commonly used in telecommunications and ***opto-electronic*** equipment.

High Bit-rate Digital Subscriber Line (HDSL): A symmetrical variant of ***Digital Subscriber Line*** (DSL) technology that can provide data rates of approximately 1.544***Mbps*** using two pairs of ***twisted-pair*** copper wire, and 2.048***Mbps*** using three pairs.

High Definition Television (HDTV): ***Highly Dispensable TV*** standard. A rare and exotic computer display option. A ***digital*** television standard that provides higher image resolution than traditional ***analog*** television broadcasts. Employs an aspect ratio of 16/9 (horizontal/vertical) and quadruple the number of picture elements or pixels of ordinary television screens (if ordinary screens had pixels). A dog technology: favored by government, always the dog's best friend.

HFC: See ***Hybrid Fiber Coax***

Home Page: The page on a ***web site*** that opens first when a site is entered. The home page typically includes links to other pages on the site, for easy navigation.

Host: The computer in a ***LAN*** or other network that is connected to all ***workstations*** via some form of communications medium, and provides useful information to the network. Within the context of the ***Internet,*** a host is a ***server*** that contains information (e.g., ***HTML pages***) that is shared amongst other Internet users. As used by Network Wizards in their Domain Name Survey, a host is a domain name that has an IP address associated with it. This would be any computer system connected to the Internet (via full- or part-time, direct or dialup connections).

HTML: See ***HyperText Markup Language***

HTML Page: A document written with the ***HyperText Markup Language*** (HTML).

HTTP: See ***HyperText Transfer Protocol***

Hybrid Fiber Coax (HFC): A communications system composed of both ***fiber-optic*** and ***coaxial*** cables. HFC is used primarily in ***CATV*** networks, and provides greater ***bandwidth*** than standard coax, making it ideal for providing ***cable modem*** services.

Hyperlink: An abbreviation for ***hypertext link.***

Hypertext Link: A portion of text on an ***HTML page*** that is linked to another page, document, or file. When selected, a hypertext link, or ***hyperlink*** for short, opens or executes the specified file, depending on the nature of the file.

HyperText Markup Language (HTML): A language used to write documents for the ***World Wide Web*** (WWW). HTML differs from standard text documents in that it contains its own commands, rather than relying on an ***application*** to define formatting.

HyperText Transfer Protocol (HTTP): The ***protocol*** used to establish ***World Wide Web*** (WWW) connections between ***servers*** and individual computers for the purpose of transferring ***HTML*** documents, image files, and various other data from one to the other.

Hz: See ***Hertz***

I

IC: See ***Integrated Circuit***

IDSL: See ***ISDN Digital Subscriber Line***

IEEE: See ***Institute of Electrical and Electronics Engineers***

IETF: See ***Internet Engineering Task Force***

IF: See ***Intermediate Frequency***

ILEC: See ***Incumbent Local Exchange Carrier***

Incumbent Local Exchange Carrier (ILEC): A local telephone service provider that is either an ***RBOC*** or was established before the Telecommunications Act of 1996.

Infrared: A portion of the ***electromagnetic spectrum,*** between visible light and microwave, used in some wireless and ***fiber-optic*** communications.

Input/Output (I/O): The transfers of data between a ***CPU*** and the various attached peripherals (i.e., the keyboard, ***mouse,*** and graphics display). For instance, when a key is struck on the keyboard, it sends an input command to the CPU. When the CPU processes that keystroke, it sends an output command to the graphics display, which makes visible the effect of the stroke that was just made.

Institute of Electrical and Electronics Engineers (IEEE): An organization whose members include engineers, scientists, and professionals from various electronics fields. The IEEE is involved in establishing standards in the computer and communications industries.

Integrated Circuit (IC): A collection of electronic circuits, that are joined together on ***semiconductor*** material in such a way as to perform a specific task. A ***microprocessor*** is an IC.

Integrated Services Digital Network (ISDN): A communications standard developed to provide digital connections to end-users over standard ***twisted-pair*** copper wire. There are two types of ISDN, Basic Rate Interface (BRI) and Primary Rate Interface (PRI). BRI utilizes two 64***Kbps*** bearer (B) channels for data transmission, and one 16***Kbps*** delta (D) channel for sending control data, and provides 128***Kbps*** service, while PRI typically utilizes 23 64***Kbps*** B channels, and one 64***Kbps*** D channel to provide various ***data rates.***

Intelligent Network: A network, such as the ***Public Switched Telephone Network*** (PSTN), that allows control of various network functions and services to be distributed, rather than being controlled by the sender.

Interexchange Carrier (IXC): A communications service provider that provides services between *Local Exchange Carriers*—long-distance phone companies (e.g., AT&T, MCI, and Sprint).

Intermediate Frequency (IF): A ***microwave*** frequency that has been reduced from its native ***frequency,*** so that it can be processed. This is necessary because a microwave cannot be fully processed in its native frequency.

International Standards Organization (ISO): An international organization that develops global standards for communications—both voice and data. The ISO's members include representatives of standards bodies worldwide.

International Telecommunications Union (ITU): The international organization, headquartered in Geneva, Switzerland, that brings together representatives from both government and private sectors to recommend standards for telephone and data communications systems.

Internet: The global ***Wide Area Network*** (WAN) that utilizes the ***TCP/IP protocol*** to connect business, government, home, organizational, and other computers. The Internet originated from ***ARPANET.*** Often referred to as "the net."

Internet Engineering Task Force (IETF): An organization that was founded in 1986 for the purpose of identifying technical problems on the ***Internet*** and proposing solutions to those problems. The IETF is composed primarily of volunteer members.

Internet Protocol (IP): The portion of the ***TCP/IP protocol*** responsible for transporting packets of data to the proper location, by tagging the packet with the ***IP address*** of the intended receiving computer.

Internet Service Provider (ISP): A business that provides ***Internet*** connections to individuals and businesses. An ISP typically provides such services as dial-up connections to the Internet, ***E-mail,*** and hosting of ***web sites*** through their server(s).

InterNIC: Internet Network Information Center. InterNIC is a cooperative effort of the National Science Foundation and Network Solutions, Inc., a subsidiary of SAIC (Science Applications International Corp.). InterNIC registers and manages all ***domain names*** within the .com, .net, .org, and .edu ***domains.***

Intranet: An ***IP***-based ***LAN*** that serves a specific company or organization, and is not accessible by the general public, as the ***Internet*** is. An intranet utilizes tools that are also utilized for Internet, such as ***browsers*** and ***HTML*** pages.

I/O: See ***Input/Output***

IP: See ***Internet Protocol***

IP Address: A 32-bit number used to identify a computer on the ***Internet,*** just as a street address would identify the physical location of a person's residence. An IP address consists of four segments—separated by periods. Each segment contains three numbers. *See* ***Domain Naming System and Uniform Resource Locator.***

IP Multicasting: A method of conserving network bandwidth in which a single file is transmitted to multiple users in a single transit, from the server to the first user, and from that user to the next, and so on. In contrast to unicast, in which each user receives a separate independent transmission, or a simultaneous broadcast to all users on the network.

IP Telephony: The transmission of voice over a ***TCP/IP*** network. IP telephony allows for phone calls to be placed via an ***Internet*** connection. Within the next decade, simply telephony.

ISDN: Incredibly Slow Decoy Network. See also, ***Integrated Services Digital Network***

ISDN Digital Subscriber Line (IDSL): A ***symmetric*** variant of ***Digital Subscriber Line*** (DSL) technology that provides data rates of approximately 128***Kbps,*** the same speed available using ***ISDN*** technology.

ISO: See ***International Standards Organization***

ISP: See ***Internet Service Provider***

ITU: See ***International Telecommunications Union***

IXC: See ***Interexchange Carrier***

J

Java: A ***programming language*** and set of ***application programming interfaces*** (APIs) created by Sun Microsystems, used to write both ***applets*** and ***applications.*** Java programming can be run on any platform (e.g., Windows, ***UNIX,*** etc.) that has a Java ***virtual machine.*** For this reason, Java has become widely used on the ***Internet*** and corporate ***intranets.*** *See also* ***Microsoft Java.***

JavaScript: A ***scripting language,*** created by Netscape, which is similar to Java, but not as powerful. JavaScript can be embedded within an ***HTML page,*** and is, therefore, supported by most browsers.

Jini: A technology developed under the leadership of Bill Joy of Sun Microsystems that allows computers and peripheral devices to establish network connections without the complex configurations and drivers required for conventional network connections. With networks increasingly running faster than the backplane buses of computers, Jini makes possible the disaggregation of computer functions across the network without serious damage to performance.

Joint Photographic Experts Group (JPEG): A standard, created by the

ISO/ITU, for the compression of graphic image files.

JPEG: See ***Joint Photographic Experts Group***

J-Script: See ***JavaScript***

K

k: See ***Kilo-***

K: See ***Kilo-***

Ka-Band: A portion of the ***electromagnetic spectrum.*** The generally accepted range includes the ***frequencies*** from 17 to 36***GHz.*** However, frequencies slightly higher or lower are sometimes referred to as Ka-band. Ka-band frequencies are commonly used for satellite communications.

Kb: See ***Kilobit***

KB: See ***Kilobyte***

Kbps: ***Kilobits*** per second. A unit of measure that is based on the amount of data, measured in kilobits, that is transferred through some form of communications medium at one time.

kHz: See ***Kilohertz***

Kilo- (k): A prefix that usually denotes 10^3 or 1,000. When used in computer terminology, kilo- is abbreviated with an uppercase K, and is actually equal to 2^{10}, or 1,024, the power of 2 that is closest to 1,000.

Kilobit (Kb): A kilobit is literally equal to 2^{10} ***bits,*** or 1,024 bits, but it is often calculated with a base of 10 (10^3), making it equal to 1,000 bits.

Kilobyte (KB): A kilobyte is literally equal to 2^{10} ***bytes,*** or 1,024 bytes, but it is often calculated with a base of 10 (10^3), making it equal to 1,000 bytes.

Kilohertz (kHz): A measurement of frequency equal to 1,000 ***hertz.***

Ku-Band: A portion of the ***electromagnetic spectrum.*** The generally accepted range of Ku-band frequencies includes 12 to 14GHz. However, frequencies slightly higher or lower are sometimes referred to as Ku-band. Ku-band frequencies are commonly used in satellite communications.

L

Lambda: A ***Wavelength Division Multiplexing*** (WDM) ***channel.***

Lambda Miles: The measurement, in miles, of individual wavelengths transmitted within a single fiber (i.e., ***fiber miles*** multiplied by ***Wavelength Division Multiplexing channels***).

LAN: See ***Local Area Network***

Latency:

1. The time it takes for a signal to travel between two points on a network.

2. The ***memory*** access delay between a ***CPU's*** request for data and the subsequent transmission of ***bits*** from memory. *See* ***Dynamic Random Access Memory.***

LEC: See ***Local Exchange Carrier***

LEO: See ***Low-Earth-Orbit Satellite***

Light Speed: The speed at which light travels—186,282 miles per second or roughly 300,000,000 meters per second.

Linux: An operating system that is based on ***Unix.*** Linus Torvalds developed Linux, with the help of volunteer programmers all over the world. Linux is distributed free of charge, and continues to be developed.

LMDS: See ***Local Multipoint Distribution Service***

LO: See ***Local Oscillators***

Local Area Network (LAN): A system of computers that are relatively close in proximity, usually within the same building, and are connected by some communications medium (i.e., ***fiber optics, twisted-pair,*** or wireless) and support the same ***protocol.*** Other devices that might be attached to a LAN include printers and storage devices.

Local Exchange Carrier (LEC): A telephone service provider that provides services to the geographic area surrounding its ***central office.*** These companies can fall into one of two categories: ***Competitive Local Exchange Carrier*** (CLEC) or ***Incumbent Local Exchange Carrier*** (ILEC).

Local Loop: The connection between an end-user's telephone or ***modem*** and the service provider's ***central office.*** Increasingly supplied by cable TV companies.

Local Multipoint Distribution Service (LMDS): Wireless two-way services that are based in the 28 ***GHz frequency*** range, and use small line-of-sight antennas to communicate. LMDS services include wireless ***Internet*** access and wireless ***cable television*** (CATV).

Local Oscillators (LO): The device used to reduce a ***microwave*** frequency to an ***intermediate frequency,*** so that it can be processed.

Low Earth Orbit Satellite (LEO): A satellite, used in various forms of communications, that orbits within five hundred miles of the Earth. Because of LEO's close proximity to the Earth, very small antennae can be used to receive its signals and delays are comparable to terrestrial ***fiber optics.*** *Contrast with* ***GEO.***

M

M: See ***Mega-***

MAE: See ***Metropolitan Area Exchange***

Major Trading Area (MTA): A geographical region of the U.S., as defined by the ***Federal Communications Commission's*** (FCC) revisions of the *Rand McNally 1992 Commercial Atlas and Marketing Guide,* based on certain economic and industrial criteria, in order to allocate areas of service. Each MTA is typically made up of several ***BTAs,*** and is named for the city that those BTAs would be most likely to use for means of commerce. There are presently 51 MTAs in the U.S.

MAN: See ***Metropolitan Area Network***

Mbone: Multicast ***backbone.*** The collection of networks on the ***Internet*** that support ***IP multicast protocol.***

Mb: See ***Megabit***

MB: See ***Megabyte***

Mega- (M): A prefix that denotes 10^6 or 1,000,000. In computer terms, however, mega- is actually equal to 2^{20}, or 1,048,576, the power of 2 that is closest to one million.

Megabit (Mb): Literally equal to 2^{20} ***bits,*** or 1,048,576 bits, but often calculated with a base of 10 (10^6), making it equal to one million bits.

Megabyte (MB): Literally equal to 2^{20} ***bytes,*** or 1,048,576 bytes, but often calculated with a base of 10 (10^6), making it equal to one million bytes.

Megahertz (MHz): 10^6 ***hertz*** or 1,000,000 hertz.

Memory: The hardware in a computer that stores data temporarily, so that the ***microprocessor*** can access it more quickly than going to the hard drive or storage device. The term *memory* is often used synonymously with ***Random Access Memory*** (RAM).

Metal Oxide Semiconductor Field-Effect Transistor (MOSFET): A type of ***field-effect transistor*** (FET) in which the ***gate*** is insulated with metal oxide.

Metropolitan Area Exchange (MAE): MAEs provide interconnect points through which ***ISPs*** transfer information to the rest of the ***Internet.*** MAEs are owned and operated by WorldCom.

Metropolitan Area Network (MAN): A high-speed network that connects computers in a larger geographic area than a ***Local Area Network,*** and yet a smaller area than that covered by a ***Wide Area Network.*** A MAN is typically the interconnection of LANs within an urban area, such as a city.

MHz: See ***Megahertz***

Micron (mm, m): A micrometer, or one thousandth of a millimeter.

Microcosm: The domains of technology unleashed by discovery of the inner structure of matter in quantum theory early in the twentieth century. The epitome of the microcosm is the microchip.

Microprocessor: A ***central processing unit*** (CPU) that is built on a single ***chip.***

Microsoft Java: A fast but defective variant of ***Java*** that can be used only on Microsoft machines.

Microwave: The portion of the ***electromagnetic spectrum,*** beginning with 1 ***GHz,*** that is used for many different wireless communications—including ***PCS,*** satellite, ***LMDS, MMDS,*** and many other services.

Million Instructions Per Second (MIPS): The measure of a computer's speed based on how many millions of instructions it can process each second.

MIME: See ***Multi-Purpose Internet Mail Extensions***

MIPS: See ***Million Instructions Per Second***

MMDS: See ***Multichannel Multipoint Distribution Service***

MMX: A variant of Intel's Pentium ***microprocessor*** that is designed to process video, audio, and graphical data more quickly and efficiently than non-MMX processors.

Modem: Modulator/Demodulator. A peripheral computer device used to convert ***digital*** signals to ***analog frequencies*** for transmission over standard phone lines, a process called ***modulation,*** and for conversion from analog frequencies to digital signals, a process called ***demodulation.*** Modems are used to establish a connection between two computers over standard phone lines.

Modulation: A physical change in a ***waveform.*** This type of change includes a modification in the ***amplitude*** or ***frequency*** in order to transmit information.

MOSFET: See ***Metal Oxide Semiconductor Field-Effect Transistor***

Motion Picture Experts Group (MPEG): An ***ISO/ITU*** standard for the compression of video images. Various levels of MPEG specifications have been developed. A number immediately following MPEG (e.g., MPEG-1) specifies each level.

Mouse: A computer peripheral used to move an on-screen ***cursor.***

MPEG: See ***Motion Picture Experts Group***

MTA: See ***Major Trading Area***

Multichannel Multipoint Distribution Service (MMDS): Wireless ***cable television.*** MMDS operates in the 2.6***GHz frequency*** range, and requires line of sight between the transmitter and receiver.

Multimedia: A term that refers to the combined use of such media as audio, text, graphics, and/or video clips to present information.

Multiplexer: A device used to merge multiple signals so that they can be sent over a single ***channel. See multiplexing.***

Multiplexing: Also known as MUXing. A method of communications by which multiple signals are transmitted through a single ***channel*** (***FDM*** and ***TDM***) or ***fiber*** (***WDM***).

Multi-Purpose Internet Mail Extensions (MIME): A method of encoding files that are not text based (i.e., audio, video, and graphical) for transmission via ***Internet E-mail.***

Multitasking: The processing of two or more ***applications*** simultaneously. The number of applications that can be processed depends on the speed of the ***Central Processing Unit,*** capacity of the ***memory,*** and the abilities of the ***operating system.***

N

N: See ***Nano-***

Nano- (n): A prefix that denotes one billionth or 10^{-9}.

Nanometer (nm): One billionth of a meter or 10^{-9} meters.

Nanoseconds (ns): One billionth of a second or 10^{-9} seconds.

NAP: See ***Network Access Point***

Narrowband: Bandwidth that is low, relative to the transmission medium. In telecommunications, a channel that is not capable of ***T-1*** data rates.

National Center for Supercomputing Applications (NCSA): A research center, located at the University of Illinois at Urbana-Champaign, that was founded in 1985 for the purpose of providing supercomputer resources to universities and organizations. NCSA was the home of the first graphical ***Internet browser,*** Mosaic.

National Science Foundation Network (NSFNet): The network that was established by the National Science Foundation for the purpose of replacing ***ARPANET*** with a network for civilian use. NSFNet was the ***backbone*** of the ***Internet*** until the Internet's commercialization in 1995.

NC: See ***Network Computer***

NCSA: See ***National Center for Supercomputing Applications***

Network Access Point (NAP): Exchange points in the ***Internet*** where ***Internet Service Providers*** interconnect to transmit data to one another. Four NAPs were established by the National Science Foundation at the time of the Internet's commercialization in 1995, and additional ones have been added subsequently.

Network Computer (NC): Also Net PC or ***thin client.*** A computer designed to be centrally managed by a server or some other fully equipped computer. An NC typically consists of a ***microprocessor*** and ***RAM,*** while a Net PC adds a hard drive, but both lack features of a stand-alone computer, or fat client, such as ***CD-ROMs,*** floppy drives, and expansion slots.

Node: A component in a network (e.g., a computer, printer, or storage device).

Nonvolatile Memory: ***Memory*** that does not require a power source in order to retain its content. Examples include ***ROMs*** and ***EPROMs.*** *Contrast with* ***volatile memory.***

NSFNet: See ***National Science Foundation Network***

O

Object Linking and Embedding (OLE): A technology, created by Microsoft, for sharing information between programs. Through the use of OLE, whenever data in a file is changed, the corresponding data will be automatically changed in linked files.

OC: See ***Optical Carrier***

OC-x (such as **OC-3, OC-12, OC-48,** etc.): See ***Optical Carrier***

OCDMA: See ***Optical Code Division Multiple Access***

OLE: See ***Object Linking and Embedding***

Open Systems Interconnection (OSI): A seven-layer model for a communications network that was established by the ***ISO,*** and provides information for ***protocols*** all the way from the user's computer interface to the transmission medium. The layers are: 7-Application, 6-Presentation, 5-Session, 4-Transport, 3-Network, 2-Datalink, and 1-Physical.

Operating System (OS): The ***software*** that integrates a computer's resources. The OS directs both ***hardware*** and software to interact. Software run on the computer must be able to communicate with the OS, and is thus made compatible with other software utilized on the system. Meanwhile, the operating system also controls the system's hardware, allocating resources for each function desired.

Optical Carrier (OC): The transmission rates set forth in the ***Synchronous Optical Network*** (SONET). The base optical carrier rate is 51.84***Mbps,*** designated OC-1. The rates increase by multiplying by the base rate. Therefore, an OC-2 would be equivalent to 51.84 times 2, or 103.68Mbps, and so on.

Optical Code Division Multiple Access (OCDMA): The principles of ***CDMA*** applied within a ***fiber-optic*** network.

Opto-Electronic: Equipment that translates electrical ***frequencies*** into optical signals, and vice versa. Opto-electronic equipment is used in ***fiber-optic*** networks.

OS: See ***Operating System***

OSI: See ***Open Systems Interconnection***

P

P: See ***Peta-***

Packet: A unit of data, in a ***packet switched*** network, that is an appropriate size for network transmission. ***Ethernet*** packets can be as large as 1,500 ***bytes,*** while the average packet on the ***Internet*** in 1998 was roughly 300 bytes. Each packet includes an identifier designating where the data is to be routed. Analogous to an envelope containing a message in a postal service dispatching hundreds of millions of letters per second (as opposed to hundreds of millions of letters per week).

Packet Switching: A method of communications in which ***packets*** are transmitted over a dedicated ***channel.*** In a packet switching network, however, the channel is dedicated to each packet only as long as it takes to transmit that packet. After each packet's transmission, the channel is assigned to a subsequent packet. Contrast with circuit switching, the system used by the ***public switched telephone network*** (PSTN), in which a full connection is maintained between two parties for the duration of the call or transmission.

Paradigm: A structure of propositions that organizes an otherwise jumbled assembly of facts. See the laws at the beginning of this glossary. The father of scientific paradigm theory was University of Chicago philosopher Thomas Kuhn, who believed that paradigms rose and fell like fashions—that scientific knowledge was cyclical rather than cumulative. However, scientific knowl-

edge manifests itself in technology, powerful and practical machines that cumulatively and empirically validate the new paradigms and generate new science in the process. Paradigms thus reflect the profound and permanent truths of the universe. Many readers, however, will prefer Harry Newton's observation that paradigm "is a word typically used by people who want to sound a little more pompous and intellectual than you and I."

PCB: See ***Printed Circuit Board.*** Also, a chemical, innocuous in all usual applications, but believed by the imaginative or litigious, in the government and out, to cause cancer, birth defects, and other disorders. Hence, a chemical that causes lawsuits.

PCM: See ***Pulse Code Modulation***

PCMCIA: People Cannot Memorize Computer Industry Acronyms. See also, ***Personal Computer Memory Card International Association.***

PCMCIA Card: Also known as a PC Card. A credit card–size device that plugs into a small slot on a notebook computer. There are numerous devices available as PCMCIA cards—the most common being network cards and ***modems.*** Each card is based on standards established by the ***Personal Computer Memory Card International Association*** (PCMCIA).

PCS: See ***Personal Communications Services***

PDA: See ***Personal Digital Assistant***

Perl: See ***Practical Extraction and Report Language***

Personal Communications Services (PCS): Wireless communications services based in the 2***GHz frequency*** range. PCS services use a portion of the spectrum that was assigned and auctioned off by the ***Federal Communications Commission*** (FCC).

Personal Computer Memory Card International Association (PCMCIA): The standards body that was created to establish standards that allow peripheral devices to be connected to portable computers. The ***PCMCIA card,*** or PC Card, is based on standards established by the PCMCIA. The acronym PCMCIA has also come to stand for People Can't Memorize Computer Industry Acronyms.

Personal Digital Assistant (PDA): Small handheld computing devices designed to provide electronic organizational tasks, including calendar and contact database functions.

Personal Handyphone System (PHS): A standard for wireless ***digital*** communications services launched by DDI, an entrepreneurial telco known as the MCI of Japan. PHS offers a cheap stripped down cellular service without efficient handoffs to neighboring cells as the user moves. In 1998, DDI endorsed

CDMA for their next system.

Peta- (P): A prefix that denotes 10^{15} or one quadrillion. In computer terms, however, peta- is actually equal to 2^{50}, or 1,125,899,906,842,624, the power of 2 that is closest to one quadrillion.

Petabits (Pb): A petabit is literally equal to 2^{50} ***bits,*** or 1,125,899,906,842,624 bits, but is often calculated with a base of 10, making it equal to 10^{15} or one quadrillion bits.

Petabytes (PB): A petabyte is literally equal to 2^{50} ***bytes,*** or 1,125,899,906,842,624 bytes, but is often calculated with a base of 10, making it equal to 10^{15} or one quadrillion bytes. ***Internet*** traffic in 1998 was measured in petabytes per week, ***LAN*** traffic in ***exabytes.***

PHS: See ***Personal Handyphone System***

Plain Old Telephone Service (POTS): Basic telephone service—including only the functions required for connection to the ***Public Switched Telephone Network*** (PSTN).

PLD: See ***Programmable Logic Device***

Plug-in: A ***software application,*** usually quite small in size, which is downloaded and added to another application, most often a ***browser,*** to support a specific function or file format. An example of this is the Adobe Acrobat Reader.

PPP: See ***Point-to-Point Protocol***

Point-of-Presence (POP): The location at which a user attains access to a ***Wide Area Network,*** such as the ***Internet.***

Point-to-Multipoint: A connection between one location and many others in a network.

Point-to-Point: A direct connection between two locations in a network.

Point-to-Point Protocol (PPP): A ***protocol*** for data transmissions over dial-up connections that was developed by the ***Internet Engineering Task Force*** (IETF) in 1991. PPP is a more secure protocol than ***Serial Line Internet Protocol*** (SLIP).

POP:

1. See ***Point-of-Presence***

2. Unit of measurement of the potential subscribers within a geographically defined telecommunications license (see ***MTA*** and ***BTA***). The number of POPs within a market is equivalent to the population of the region.

Post, Telephone, and Telegraph (PTT): The agency, typically found in European countries, that is responsible for retarding the provision of postal and telecommunications services. PTTs is usually administered by the national government in whose territory they operate.

POTS: See ***Plain Old Telephone Service***

Power/Delay Product: The product of the switching speed of a device (delay) and its power usage (heat). An index of the efficiency of a ***semiconductor*** device.

Practical Extraction and Report Language (Perl): A ***scripting language,*** developed by Larry Wall at NASA's Jet Propulsion Laboratory, which uses combined commands from ***C*** and ***Unix.*** Perl is used to write ***Common Gateway Interface*** (CGI) ***scripts,*** along with a variety of other scripts for ***Internet*** applications.

Printed Circuit Board (PCB): A fiberglass or plastic board on which copper pathways are printed to connect holes in which various electronic components are mounted. PCBs are used to create various add-in cards (e.g., ***memory,*** internal ***modems***) for computers.

Processor: See ***Central Processing Unit***

Programmable Logic Device (PLD): A ***chip*** that can be programmed at the customer's location, rather than during the manufacturing process.

Programmable Read-Only Memory (PROM): ***Memory*** that differs from ***ROM*** in that it can be programmed after the manufacturing process, using a special programmer. A PROM differs from an ***EPROM*** in that it cannot be reprogrammed.

Programming Language: A language used to write instructions for a computer. A set of instructions makes up an ***application,*** or program. Programming languages are like human languages in that certain grammar and syntax must be learned in order to read, write, or speak the language. Programming languages include ***C, C++, COBOL,*** and ***Java.***

PROM: See ***Programmable Read-Only Memory***

Protocol: A set of recognized rules. In the computer industry, protocols are established to facilitate intelligible transmission of data among users.

Proxy Server: A ***server*** that works as part of a ***firewall*** by eliminating direct connections between one network and another. All traffic is routed through the proxy server to the proper network connection. A proxy server can also be used to speed up retrieval of commonly used files (e.g., ***HTML pages*** that are accessed frequently) by storing them locally. A proxy server is also known as simply a proxy.

PSTN: See ***Public Switched Telephone Network***

PTT: See ***Post, Telephone, and Telegraph***

Public Switched Telephone Network (PSTN): The global network used for voice telephone communications. *Contrast with* ***Dumb Network.***

Pulse Code Modulation (PCM): A method of converting ***analog*** signals into ***digital*** signals, accomplished by sampling, or measuring, the analog signal many times per second, and converting the sample into a digital signal.

Pump Lasers: Lasers that are built into ***erbium-doped fiber amplifiers*** (EDFAs) for the purpose of exciting the erbium in the fiber (i.e., to power the amplifier).

R

Radio Frequency (RF): The portion of the ***electromagnetic spectrum*** that is between the audible and visible range of frequencies.

RADSL: See ***Rate-Adaptive Digital Subscriber Line***

RAID: See ***Redundant Array of Independent Disks***

RAM: See ***Random Access Memory***

Random Access Memory (RAM): A type of ***memory*** that can be read and written to. RAM is the standard memory used in computers to avoid the ***CPU*** sending requests to the hard drive as often. This increases the speed at which the CPU can access data.

Rate-Adaptive Digital Subscriber Line (RADSL): An ***asymmetrical*** variant of ***Digital Subscriber Line*** (DSL) technology that can provide up to approximately 1***Mbps upstream*** and 7***Mbps downstream.***

RBOC: See ***Regional Bell Operating Company***

Read-Only Memory (ROM): ***Memory*** that is programmed at some point during the manufacturing process, and can be read by a computer, but not modified.

Real Time: Refers to the ability of a computer to process and respond to data in synchronization with an event as it occurs.

Real-Time Operating System (RTOS): An ***operating system*** that can provide an instant response to input data.

Reduced Instruction Set Computing (RISC): A ***microprocessor,*** such as the ***SPARC,*** that is programmed to focus its resources on completing simple tasks very quickly. This processor is much faster at completing a small number of tasks than a ***CISC*** processor, but falls short of CISC speeds when completing complex tasks.

Redundant Array of Independent Disks (RAID): A method of increasing network performance and reliability, by distributing stored data among several hard disks that are managed by specialized software.

Regional Bell Operating Company (RBOC): One of the seven companies—Ameritech, Bell Atlantic, BellSouth, NYNEX, Pacific Bell, Southwestern Bell, and US West—formed in the divestiture of AT&T in 1983. Each RBOC owned two or more Bell Operating Companies (BOCs) through which local telephone service was provided.

RF: See ***Radio Frequency***

RISC: See ***Reduced Instruction Set Computing***

ROM: See ***Read-only Memory***

Router: A device in a ***LAN*** or ***WAN*** that determines the best available route to send a data ***packet*** in order to most quickly and efficiently get it to its destination.

Route Miles: The measurement, in miles, of distinct routes in a ***fiber-optic*** system. One route mile might contain 96 ***cable miles*** per conduit and scores of thousands of ***fiber miles*** (each of which could contain as many as 120 separate wavelengths or ***lamda miles***).

RTOS: See ***Real-Time Operating System***

Runtime: The period in time when an ***application*** is actually being executed.

S

Scalar Processor Architecture (SPARC): A ***reduced instruction set computing*** (RISC) ***microprocessor*** produced by Sun Microsystems.

S-CDMA: See ***Synchronous Code Division Multiple Access***

Scripting Language: A language that typically contains fewer commands and functions than a ***programming language,*** and is used to write routine commands that add functionality to an ***application.*** A script is processed by the application that it resides in, rather than standing alone. Some examples of scripting languages include ***JavaScript*** and ***Perl,*** which are used to add forms and automated functions to ***HTML pages.***

SDRAM: See ***Synchronous Dynamic Random Access Memory***

SDSL: See ***Single-line Digital Subscriber Line***

Search Engine: A program that searches the ***Internet, World Wide Web,*** newsgroups, and ***FTP*** sites based on user defined information, such as keywords.

Some search engines include Yahoo, AltaVista, and Excite. Some ***web sites*** include their own search engines that are limited to searching that particular site.

Semiconductor: A substance that is neither a good conductor, nor a good insulator, i.e., a substance that is a conductor when an electrical charge is passed through it, or a nonconductor when the charge is taken away. "Semiconductor" has also come to refer to electronics components that are made of materials that include ***silicon, silicon germanium,*** and ***gallium arsenide.***

Serial Line Internet Protocol (SLIP): A ***protocol*** for data transmissions over dial-up connections.

Server: Computer ***hardware*** and/or ***software*** on a network that controls such functions as network access, file sharing, print managing, and central ***application*** processing. The server is the command center of a network. A network can contain more than one server. *See also* ***Client/Server.***

Set-Top Box: An appliance that is attached to a television, and used to provide such services as interactive ***cable television*** or ***Internet*** and ***E-mail*** access.

SGML: See ***Standard Generalized Markup Language***

Si: See ***Silicon***

SiGe: See ***Silicon Germanium***

Silica: The compound SiO^2. Silica is found throughout many minerals, such as quartz, and is used in the production of glass.

Silicon (Si): The most commonly used base element in the production of ***integrated circuits.*** Silicon is one of the most abundant elements on Earth, second only to oxygen.

Silicon Germanium (SiGe): A compound used to produce various types of ***integrated circuits.*** SiGe was originally developed by a team of IBM researchers led by Bernard Meyerson, and competes with ***gallium arsenide.*** Because the process of combining silicon and germanium is very difficult, a technique called ***UHV/CVD*** was developed.

SIMM: See ***Single In-Line Memory Module***

Simple Mail Transfer Protocol (SMTP): The standard ***protocol*** for sending ***email*** messages.

Single In-Line Memory Module (SIMM): A small ***printed circuit board*** (PCB), on which ***RAM chips*** are mounted, that is equipped with connections that allow it to be plugged into sockets on a computer's motherboard. SIMMs provide a method for adding ***memory*** to a computer. SIMMs are predecessors of ***DIMMs.***

Single-line Digital Subscriber Line (SDSL): A ***symmetric*** variant of ***Digital***

Subscriber Line (DSL) technology that can provide data rates up to 2***Mbps.***

SLIP: See ***Serial Line Internet Protocol***

Smart Card: A credit card–like plastic card with an embedded microchip, capable of storing and/or processing data. Smart cards are capable of storing E-Cash, electronic tickets, and personalized data, such as network or computer log-on information or medical records.

Smart Radio: Also referred to as a ***software*** radio. An advanced radio in which ***frequencies,*** tuning, and ***protocols*** are defined using software programming, rather than fixed ***hardware.*** Such a radio might be used for a wireless handset or ***base station,*** giving it the capability of utilizing various previously incompatible protocols interchangeably, such as ***AMPS, GSM,*** and ***CDMA.***

SMR: See ***Specialized Mobile Radio***

SMTP: See ***Simple Mail Transfer Protocol***

Software: A set of instructions that directs the ***hardware*** in the processing of data. Both ***operating systems*** and ***applications*** are considered to be software. *Also referred to as program.*

Software Radio: See ***Smart Radio***

SONET: See ***Synchronous Optical Network***

SPARC: See ***Scalar Processor Architecture***

Specialized Mobile Radio (SMR): Mobile radios that are used by various businesses, such as taxis and ambulance services. SMR ***frequencies*** are also used by carriers, such as Nextel, to provide ***cellular*** like service to consumers. These radios require licensing through the ***Federal Communications Commission*** (FCC).

Spectronics: Technology seen from the point of view of the electromagnetic spectrum. Examples include 60 ***hertz*** power lines, 400 ***megahertz microprocessors,*** and 190 ***terahertz fiber optics.***

Spherical Semiconductor: An ***integrated circuit,*** developed by BALL Semiconductor, created on the surface of a one-millimeter ***silicon*** sphere, and designed to compete with conventional integrated circuits.

Spider: A program that continually searches the ***World Wide Web*** (WWW) for new hyperlinks and ***HTML pages.*** A spider collects and indexes its findings for later access by a ***search engine.***

Spread Spectrum:

1. A form of radio transmission in which the data is divided into ***packets*** that are each assigned an identifier. These packets are then spread across the available ***bandwidth*** for transmission, and regrouped at the receiving end.

2. A form of radio transmission whereby the ***frequency*** used changes continually. In conventional radio transmission the signal is broadcast on one frequency that remains constant. A spread spectrum signal continuously jumps to multiple subsequent frequencies during the transmission. This form of spread spectrum is referred to as "frequency hopping."

SRAM: See ***Static Random Access Memory***

Standard Generalized Markup Language (SGML): A standard, as defined by the ***International Standards Organization*** (ISO), for defining formatting within a text document. One of the best known SGMLs is ***HyperText Markup Language.***

Static Random Access Memory (SRAM): A type of ***Random Access Memory*** that requires power in order to retain the information being stored in it. SRAMs typically have much faster access speeds than ***DRAMs.***

Stream: A steady flow of data ***bits.*** Also called a ***bit stream.***

Stupid Network: See ***Dumb Network***

Switch:

1. A circuit that has two positions—on and off. The switch is controlled using an electrical signal.

2. A computer, or electrical device, through which data network and/or telecommunications signals are routed to their proper destinations.

Switched Network: See ***Public Switched Telephone Network***

Symmetric: Used to describe communications technologies in which the ***upstream*** and ***downstream data rates*** are identical—e.g., ***High Bit-rate Digital Subscriber Line.***

Synchronous Code Division Multiple Access (S-CDMA): A form of ***Code Division Multiple Access*** (CDMA), developed by Terayon, for use in providing cable modem services. S-CDMA transmits signals through ***Hybrid Fiber Coax*** using the principles of CDMA, thereby increasing the quality of the ***upstream.***

Synchronous Dynamic Random Access Memory (SDRAM): A type of ***DRAM*** that has very fast access rates, achieved by preparing the next memory address to be read as the previous one is still being accessed.

Synchronous Optical Network (SONET): A high-speed fiber-optic networking system. SONET facilitates speeds ranging anywhere from 51***Mbps*** up to many ***gigabits*** per second, depending on the size of the ***optical carrier.***

T

T: See ***Tera-***

T-1: A digital communications line that can facilitate ***data rates*** of up to 1.544***Mbps.*** T-1 lines are commonly used for high-speed ***Internet*** connections, typically by institutional entities.

T-3: A digital communications line that can facilitate ***data rates*** of up to 44.736***Mbps.***

TACS: See ***Total Access Communications System***

TCP/IP: See ***Transmission Control Protocol/Internet Protocol***

TDM: See ***Time Division Multiplexing***

TDMA: See ***Time Division Multiple Access***

Telecosm: The domains of technology unleashed by the discovery of the ***electromagnetic spectrum*** and the photon. ***Fiber optics, cellular telephony,*** and satellite communications are examples.

Telephony: The technology and equipment used to provide telephone services—including the conversion of sound to electrical or optical signals, the transmission of those signals from one point to another, and the reconversion to sound.

Tera- (T): A prefix that denotes 10^{12} or one trillion. In computer terms, however, tera- is actually equal to 2^{40}, or 1,099,511,627,776, the power of 2 that is closest to one trillion.

Terabit (Tb): A terabit is literally equal to 2^{40} ***bits,*** or 1,099,511,627,776 bits, but is often calculated with a base of 10 (10^{12}), making it equal to one trillion bits.

Terabyte (TB): A terabyte is literally equal to 2^{40} ***bytes,*** or 1,099,511,627,776 bytes, but is often calculated with a base of 10 (10^{12}), making it equal to one trillion bytes.

Terahertz (THz): 10^{12} ***hertz*** or one trillion hertz.

Terapops: A ***point of presence*** (POP) with a capacity of terabits per second.

Thin Client: See ***Network Computer***

Time Division Multiple Access (TDMA): A technology used in satellite and ***cellular*** communications whereby each ***channel*** in the system is split into time slots—typically three per channel. Transmissions are then split into ***packets*** that are inserted into each slot, enabling three channels over one. *Contrast with*

Code Division Multiple Access *and* ***Frequency Division Multiple Access.***

Time Division Multiplexing (TDM): A form of transmission by which a single channel can be sliced into time packets. By this method, several channels are melded together in sequence prior to transmission, and then separated into their original signals upon reception. *Contrast with* ***FDM.***

Token Ring: A networking ***protocol*** that was developed by IBM, and provides specifications for networking up to 255 workstations, using ***twisted-pair*** or ***coaxial*** wiring. Functionally superior to ***Ethernet,*** which outsold Token Ring by five to one because of greater simplicity and decentralization.

Total Access Communications System (TACS): An ***analog cellular*** communications system that was developed by Ericsson.

Transistor: Originally, transfer resistor. A small device, made up of ***semiconductor*** material, that is used to regulate electrical current, by acting as a ***gate,*** or ***switch.*** An ***integrated circuit*** (IC) is composed of interconnected transistors and other devices.

Transmission Control Protocol/Internet Protocol (TCP/IP): A communications protocol used in the ***Internet.*** The TCP portion of the ***protocol*** manages data being sent into ***packets,*** while the IP portion contains the addressing information that designates their proper destination.

Transponder: A transmitter and responder. Transponders are the portion of a satellite that receive signals from transmitters on earth, and transmit them to ground receivers via a different ***frequency.***

Trunk: A connection between two switching stations on a network. A trunk typically carries large amounts of traffic, as in the connections between ***NAPs*** or ***MAEs*** on the ***Internet.***

Twisted-Pair: A type of wire that is used in traditional telephone networks. A twisted-pair is a set of wires—usually copper—that are individually insulated, and then twisted together.

U

UHV/CVD: See ***Ultra-high Vacuum/Chemical Vapor Deposition***

Ultra-high Vacuum/Chemical Vapor Deposition (UHV/CVD): The process, developed by an IBM research team led by Bernard Meyerson, by which germanium is added to ***silicon*** to form ***silicon germanium*** (SiGe).

Uniform Resource Locator (URL): The address that can be typed into a ***browser*** to find a specific file or resource on the ***Internet.*** URLs can include the ***protocol*** used (e.g., ***HTTP*** or ***FTP***), the name of the server on which the file resides (e.g., www.gildertech.com), and oftentimes the name of the file or resource (e.g., http://www.gildertech.com/index.html).

Unix: An ***operating system*** that was developed at AT&T in 1969, and further developed by UC Berkeley, Sun, SCO, and IBM, among others. Unix is freely available to academic institutions and government agencies, and is recognized as the standard for networking ***servers*** and ***workstations.***

Upstream: The path, in communications, through which the end-user transmits data to the service provider. *Contrast with* ***downstream.***

URL: See ***Uniform Resource Locator***

V

VBI: See ***Vertical Blanking Interval***

VDSL: See ***Very High Asynchronous Digital Subscriber Line***

Vertical Blanking Interval (VBI): The interval of time, in television picture formation, during which the ***electron*** beam is turned off while the electron gun moves back to the right-hand side of the screen before beginning the next line. This time period is used to transmit signals for such services as Intel's Intercast and closed captioning.

Very High Asynchronous Digital Subscriber Line (VDSL): An ***asymmetric*** variant of ***Digital Subscriber Line*** (DSL) technology that can provide rates up to 2.3***Mbps upstream,*** and 51.84***Mbps downstream.***

Very Large-Scale Integration (VLSI): A ***chip*** that contains more than one hundred thousand ***transistors.***

Video-on-Demand (VoD): An information technology that allows video files, such as movies, to be delivered via a network connection to an end-user's PC or TV, and offers the end-user "VCR-like" functionality (i.e., stop, rewind, and fast forward), on-demand access, and full-screen, full-motion viewing. With ***microprocessors*** running at a ***gigahertz*** and linked to networks running at a ***gigabit*** per second, such functions will be performable on any PC linked to the Internet.

Video Random Access Memory (VRAM): A type of ***RAM,*** for video applications, VRAM provides connections to both the ***CPU*** and the computer's video circuitry, which provides much faster performance than standard ***DRAM*** can provide.

Virtual Machine: A ***software*** interpreter that translates code written in a language unknown to a ***CPU*** into commands the CPU is able to recognize. The best known virtual machine is the ***Java*** interpreter or Java runtime engine that translates Java code line by line in real time, into the machine language of a particular computer.

Virtual Private Network (VPN): A network that is built on public infrastructure, such as the ***Internet,*** but provides private connections through the use of data encryption.

Visible Light: Wavelengths between 400 and 700 ***nanometers*** and frequencies between 430 and 750 ***terahertz.*** The portion of the ***electromagnetic spectrum*** that is visible to the human eye and is not widely deemed by influential phobics to cause cancer. A one over 10,000,000,000,000,000,000,000,000 part of the total spectrum.

VLSI: See ***Very Large-Scale Integration***

VoD: See ***Video-on-Demand***

Voice Grade Channel: A communications channel, approximately 4***kHz,*** that is capable of transmitting a human voice signal.

Volatile Memory: ***Memory*** that requires a power source in order to retain its content, e.g., ***RAM.*** Volatile memory loses its content when its power supply is interrupted. *Contrast with* ***nonvolatile memory.***

VPN: See ***Virtual Private Network***

VRAM: See ***Video Random Access Memory***

W

WAN: See ***Wide Area Network***

Waveform: A visual representation of a wave's characteristics plotted against time. *See example below.* One waveform is one complete cycle.

Wavelength:

1. The distance between a point on an electromagnetic wave and the identical point on the next wave in the cycle.

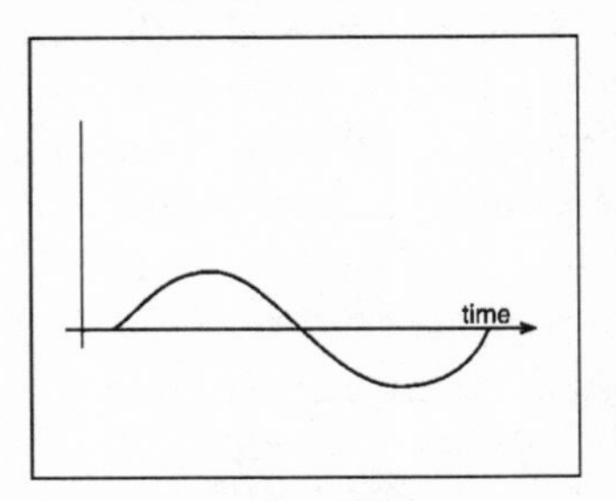

2. An individual ***channel*** defined by light ***frequency*** within a ***Wavelength Division Multiplexing*** (WDM) system.

Wavelength Division Multiplexing (WDM): A method of increasing the ***bandwidth*** of a ***fiber-optic*** network by transmitting multiple signals through each strand of fiber. This is accomplished by assigning each signal a unique light ***frequency*** or ***wavelength,*** and then transmitting multiple frequencies simultaneously. Applied to optical communications, it is another word for ***Frequency Division Multiplexing*** (FDM).

WDM: See ***Wavelength Division Multiplexing***

Web Site: A presence on the ***World Wide Web.*** A web site can include either single or multiple ***HTML pages.*** Multiple pages on a site would typically be linked using ***hyperlinks.***

Wide Area Network (WAN): A network of computers—typically spread over a large geographic area. Each computer on the WAN must be connected to the others via a communications medium (i.e., ***fiber optics, twisted-pair,*** or wireless) and use compatible ***protocols.*** The ***Internet*** is an example of a WAN.

Wideband: *See* ***Broadband***

Workstation: A computer that is attached to a network, and interacts with the ***server.***

World Wide Web (WWW): The worldwide network of computers that interact on the ***Internet*** using the ***HyperText Transfer Protocol.*** The WWW was conceived at the ***Conséil Europeen pour la Recherche Nucléaire*** (CERN) in 1989 by Tim Berners-Lee, for the purpose of exchanging information on nuclear research. It spread.

WWW: See ***World Wide Web***

xDSL: An acronym used to refer collectively to the various types of ***Digital Subscriber Line*** (DSL) technology. A telephone company expression that means: "We don't have any idea what kind of digital subscriber line can save our tail and tariffs against ***cable modems.***"

Index

Acknowledgments

Acknowledgment and profuse thanks are owed to editors Richard Vigilante, Rich Karlgaard, Spencer Reiss, Charlotte Allen, Andy Kessler, Joshua Gilder, Sandy Philp, and finally successful against all odds, Bruce Nichols; and to Toby Casey for work on the glossary; Bret Swanson for work on the "nine," David Minor for shaping the "list," Chuck Frank for reviving the project, and Debi Kennedy for keeping me on track. Thanks also to Bruce Chapman and the Discovery Institute for constant help, encouragement, and guidance. And most of all, to Nini, who kept me alive and in love.

—George Gilder,
aboard the Crystal Symphony,
Copenhagen, 25 June 00

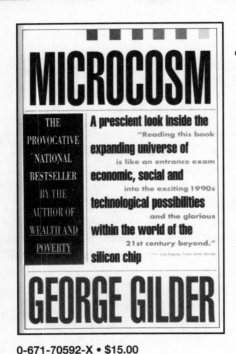